ENERGY IS LIFE

For my parents, who brought us here

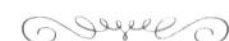

For my daughters, who will go beyond

Praise for *Energy is Life*

The book is insightful at many levels, from the fundamental nature of our existence to the political, economic, and ecological challenges we now face. In this vivid book, Zion Lights offers strong evidence, clear thinking, and good sense on the biggest issue of our time.

Steven Pinker, Johnstone Professor of Psychology, Harvard University, and the author of 12 books, including *When Everyone Knows That Everyone Knows...*

* * *

Yes, nuclear energy is dangerous. But at this point, not using it is even more dangerous. In this well-argued book, Zion Lights makes a passionate, personal plea to not forget that without energy, we are nothing. Highly recommended.

Sabine Hossenfelder, physicist and author of *Existential Physics: A Scientist's Guide to Life's Biggest Questions*

* * *

Zion Lights humanizes nuclear energy and persuasively describes the great potential of nuclear power to improve our lives.

Professor James E. Hansen, climate scientist and former Director of NASA Goddard Institute for Space Studies

* * *

The entire history of civilization can be reduced to two words: energy capture – in the teeth of entropy, humans need energy to survive and flourish. Zion Lights' Energy is Life *is the definitive case for why environmentalism must go nuclear. The fate of humanity depends on it, and this is the book that outlines how we can get there.*

Michael Shermer is the Publisher of *Skeptic* magazine and author of *The Moral Arc, Conspiracy, and Truth*

* * *

I didn't know what to expect going in, but I was really impressed. This felt like when I read The Demon-Haunted World *many many years ago… A deeply humanistic call for nuclear power from one of today's leading science communicators.*

Yishan Wong, CEO and Founder of Terraformation, a company dedicated to accelerating global reforestation, and former CEO of Reddit

* * *

In this book, Zion Lights offers the reader a masterclass in how to update your beliefs in the face of scientific evidence, how to learn from past errors, and how to engage in authentic science communication about one of the greatest challenges of our time.

Sander van der Linden, Professor of Social Psychology, University of Cambridge and Author of *Foolproof: Why We Fall for Misinformation and How to Build Immunity*

* * *

A moving and engaging account of the author's journey from anti-nuclear protestor to climate activist who understands that nuclear power is the most important technology we have for mitigating climate change.

Kerry Emmanuel, climate scientist and Professor post-tenure of atmospheric science, MIT

* * *

Energy has determined human history and determines our future too. In this powerful book, Zion Lights explains how we can produce enough of it without cooking the planet – and why nuclear power is an indispensable tool in meeting this challenge. This is a must-read for anyone seeking science-based, well-written guidance on how to build a sustainable future for all of humanity.

Atte Harjanne, Member of Parliament for the Green Party, Parliament of Finland

* * *

This book is simultaneously emotionally compelling and underpinned by scientific numeracy. This combination of fact and passion comes together to weave a hopeful vision for how nuclear energy can enable a cleaner, more prosperous, equitable and sustainable future for the world. A stellar demonstration of effective science and energy communication.

Katy Huff, Associate Professor and former Assistant Secretary at US Dept. of Energy

* * *

Lights sees things as they are, not as wealthy people imagine them. Her writing is fresh and vivid. Energy is life. I recommend this book highly.

Meredith Angwin, former chemist and author of *Shorting the Grid*

* * *

Aldous Huxley said that the richest non-fiction writing lives in three worlds at once: the intimate world of personal experience, the solid ground of fact, and the higher space of universal ideals. Lights' work moves effortlessly among these realms, a true Huxleyan trinity.

Jesse Freeston, award-winning documentary filmmaker

* * *

A call to action against climate change using science and storytelling, this is a must-read for environmental activists and policy makers alike. Zion shows how we can leave behind shibboleths and sentiment to give a clear-eyed way forward to a better future for all.

Lizzi Collinge, British Labour Party politician serving as Member of Parliament

* * *

Written with the empathy and communication skills of a master. Zion Lights weaves a compelling personal narrative with rigorous reasoning to show why abundant, reliable electricity is a prerequisite for dignity and opportunity and why nuclear energy belongs at the center of any honest climate strategy. As someone who teaches and does research in radiation science and risk communication, I'm impressed by the book's empirical treatment of radiation, accidents, and waste, and by its empathy for communities most harmed by energy poverty. This is the kind of bridge building environmentalism we need: principled, pro human, and grounded in evidence.

Professor Robert B. Hayes, PhD, CHP, PE, Nuclear Engineering Department, North Carolina State University

* * *

Energy is Life is a powerful reminder that the stories we tell shape the world we build. I was deeply struck by Zion's insistence that progress cannot come from narratives rooted in fear or deprivation, but from those that champion abundance, agency, and equity. Her advocacy for communities across India, Africa and other emerging regions gives voice to those whose lives are too often defined by other people's ideas. This is the kind of reframing our world urgently needs.

Natasha Mudhar, global impact leader who launched India's SDGs, Founder of *The World We Want Studios*

* * *

Energy is Life is not only a clear, well-grounded account of what we actually know about the energy transition and nuclear power – it's also a story of a climate movement learning to think with greater maturity, shaped by experience rather than ideology, told by one of its own. This is a book we urgently need today: comprehensive, rational, and free of dogma – yet also brave, candid, and deeply personal.

Jakub Wiech, Polish energy analyst and journalist

* * *

Zion Lights' book about energy describes how she has completely changed her views on nuclear and the environment. Her Pauline conversion from leading critic to vocal protagonist provides remarkably interesting and reassuring reading for anyone worried about the sources of energy we need.

Professor Wade Allison, Emeritus Professor of Physics and Fellow of Keble College, University of Oxford, UK

* * *

The insights in this book into the pace and scale of the transformation needed are really thought provoking and the writing shows a great depth of analysis and a personal journey which are unique. This deeply heartfelt book is essential reading for anyone interested in, let alone creating, 21st century energy policy.

Tim Stone CBE, former Chair of the Nuclear Industry Association

* * *

Perhaps the time has come for the green movement to re-evaluate its position on civilian nuclear power. In this book, Zion Lights with her impressive environmental track record and lucid science communication shows why.

Professor Alastair McIntosh, author of *Riders on the Storm: the Climate Crisis* and the *Survival of Being*

* * *

Charming and engaging… A philosophical case for pronuclear environmentalism.

Joshua S. Goldstein, co-writer with Oliver Stone of the documentary *Nuclear Now*

* * *

ENERGY IS LIFE

WHY ENVIRONMENTALISM WENT NUCLEAR

ZION LIGHTS

UNICORN

First published by Unicorn
an imprint of Unicorn Publishing Group, 2026
Charleston Studio
Meadow Business Centre
Lewes BN8 5RW

www.unicornpublishing.org

ISBN 978-1-917458-45-0

Cover design by Unicorn
Typeset by Vivian Head

Printed by Bell & Bain Ltd, Glasgow

CONTENTS

INTRODUCTION
ENERGY IS LIFE AND WHY ENVIRONMENTALISM WENT NUCLEAR

What does a saucepan have to do with a nuclear power plant? Let me tell you a story.

Childhood is a strange and wonderful thing, and the memories it leaves behind are just as curious – sometimes vivid, other times elusive. I recall our family holidays to the Welsh coast, not for the scenery or the games on the beach, but for the taste of ham sandwiches that were gritty with sand. They tasted horrible, but we were too busy having fun on the beach to care. My parents somehow stayed cheerful through those caravan trips, even when the weather was grey, wet and windy. Inevitably, sand found its way into everything – shoes, socks, bags and coat pockets – turning up for weeks after we got back to the city. While I know I had enjoyed these holidays, most of the other happy moments from those trips have slipped quietly out of reach, and all that remains is the taste of the sandy sandwich.

Some memories linger because, as children, we can't quite make sense of them. They sit in a dusty corner of the mind, waiting until we're older so we can attempt to finally dust them off, unravel them, and see where they belong. This is one of those memories. My parents were factory workers for their entire adult lives until they retired. My father stayed loyal to the same company for decades, until it went out of business, and my mother moved between factories over the years. All of her jobs demanded hard, physical work, including operating machines, packing and lifting heavy boxes, and she went from sewing garments in one place to making pots and pans in another, always standing on her feet for several hours a day, with few breaks.

At ten years old, wrapped up in my own small dramas and concerns, I didn't pay much attention to what my parents did for work. But I do have a vivid memory of my mother bringing home a stainless steel saucepan, and our family gathering around it as if it was a rare and precious jewel. I remember clearly the smell of the freshly unwrapped pan, mum lifting it carefully from its nest of paper in a large box for all to admire in our small kitchen. I didn't know quite what I felt at the time – only that we were looking at more than just a saucepan. The awe I saw in my parents' faces was contagious, the wonder palpable. Still, I sensed I didn't fully understand the importance of the moment or the pan, so I tucked the memory away, returning to it over the years; a puzzle I couldn't yet solve.

That saucepan represented a lot. On the surface, my parents had arrived in Britain with few possessions. They had worked hard to be able to own such a high quality saucepan, and this was a pan that my mother had helped to make with her own hands. Also, by doing so she was able to help feed her family. I understood that she was proud, and also that these pans were expensive. We could only afford one because she was given an employee discount since she'd been working in this factory for so long. But I still felt that I was missing the real meaning of unpacking the saucepan, and the memory sat in a corner of my mind. I revisited it as an adult many times, trying to make sense of the love of an item while being surrounded by environmentalists who believed and regularly told me that consumerism and materialism were ugly, deadly and unnecessary. How did these pieces of the puzzle fit together? And why did they matter?

Many years later, understanding dawned on me as if it had always been there, like a door quietly swinging open to a room I'd been standing outside all my life that had never truly been locked. My parents were not overcome by unwrapping a steel saucepan at home because it represented capitalism, or because they were materialistic people. They were emotional because this saucepan represented agency. First, the obvious – that my mother could afford such an expensive item, which, for a child who grew up in rural poverty with few possessions, had previously seemed impossible. Next, she could finally afford something that she wanted; an item she'd no doubt made for thousands of other people at work, but had, until now, been out of her grasp. I'm sure we can all relate to this to some degree – think about the first time you were able to afford something that you wanted for many years; something that became a precious possession that meant more than the currency you spent on it. But still, there's more – the saucepan also meant liberation. For both of my parents, but especially my mother, it meant that they had independence. The fact that she had earned this item, and she would feed her family through cooking with it for many years, was unthinkable for a woman back home, in a culture that stifles woman – and she was the first of a large family to taste these freedoms. The men in her family back home in rural India were rice farmers and had been for generations. The women did chores, got married, bore children and raised children. But now, my mother had a job, a living wage and job security. And, finally, she had freedom, because this pan meant that she had been freed her from the shackles of cooking over a crude stove fed by wood and charcoal for the rest of her life, which is how she had been taught to cook as a child in India.

If you have a saucepan, that means you have a stove with a hob that produces low fumes. Either it provides fire using gas, or, even better, it's electric and therefore fume-free. Either way, you no longer have to slave over a smoky fire with a scarf wrapped around your mouth to cook dinner. Often cooking chapattis with your bare hands directly in the fire (I tried to learn this once and do not recommend it) in relentless heat. It might not seem like a lot to us, because we are energy-rich, with cleaner and safer ways of cooking that don't harm our lungs and sting our eyes. But look through the eyes of my mother for a moment, and perhaps you can see a different picture.

In Indian culture, tea is typically made in a saucepan, and to this day my parents still make *chai* this way. They do own an electric kettle, but making tea in a saucepan is ritualistic and comforting – the slow process of measuring out spices, adding one ingredient at a time, watching the water come to a boil, then bringing it to a simmer, sieving the piping hot liquid into mugs. This was done three times a day as a way to break up the day, to pause and breathe. That said, I do think *chai* made this way tastes better than when made with an electric kettle, as the cloves and cardamom release more flavour on the hob – but I digress. Imagine your life without an electric kettle, and that is life for millions of people who do not own a saucepan. Or, they do own a saucepan, but it's low quality. The saucepan my mother brought home that day was considered to be high quality, but also it was a much more versatile tool than a kettle. While an electric kettle requires electricity, a saucepan will work with any fire or stove. For my parents, this saucepan represented the ability to make tea, cook a meal and boil water for a bath (something we did whenever our old boiler stopped working) – all things that point to freedom.

So what does a saucepan have to do with a nuclear power plant? It's entire function depends on the existence of energy. Neither a kettle nor a saucepan is truly functional without an energy source – whether it's a plug in the wall or the heat of a simple flame. Today, my parents own a significant collection of pans, which all serve different purposes in the kitchen. For them, the world changed not just when they left their home country to start a new life, but when they were able to buy a gas cooker and collect enough pans that they could cook on every section of the hob. That is why the first high quality saucepan my mother brought home from work was treated like a gift from the gods.

When I finally understood what the memory meant, and how that saucepan meant freedom, I coined a phrase to represent it: 'energy is life'. We forget this, because we are accustomed to having access to reliable electricity, with no real limit to how many appliances we plug in at home. But we also know it to be true – since I started talking about energy this way in 2020, many key figures have repeated this phrase, including the US Secretary of Energy Chris Wright[1], and a tech billionaire[2]. We tend to list our basic needs as food, water and shelter, while taking energy – which is involved in all of these needs – for granted. Our basic needs are far harder to obtain and to keep without energy.

We are energy-rich people – among the richest people on the planet. Because of this, we take energy for granted. Most of us never wonder if the lights will turn on when we flick a switch, or worry that we'll freeze in winter. We are fortunate that we don't have to fear energy poverty or endure the struggles that, for centuries, women in particular have faced to provide food, shelter and water without safe, reliable and abundant power. Our lives are built on, and intertwined with, energy, whether we realise it or not.

My mother is just one woman, but she represents millions of women who have been freed from the grip of energy poverty. For them, this freedom is not something they see only in their pay cheque or commute to work, but in the concrete, visible product of their toils: a tool made using energy, and which only works with energy. An item that means they have escaped energy poverty, and have even, in some cases, become energy-rich.

As a child I attended *gurdwara* (temple) many times with my parents, where I was offered sweet treats of *gelabi* and *gulaab jaman* and ran around barefoot with other children while the elders prayed. But the first time I saw true prayer was not in the *gurdwara*. It was when my parents unpacked and showed us that saucepan. The reverence displayed in the careful way she handled it and passed it around. The feel of the cold grey metal in my hands, the pride in her eyes, the unspoken feelings. Then later, the pride of place that saucepan was given in the kitchen, which would, over time, become populated with many other saucepans. We didn't have a *mandir* at home, but the saucepans in our kitchen were, in a way, my mother's shrine.

When I left home, mum gave me a huge saucepan. It's so big that it takes up three hobs on the stove, and I originally used it for jam making, but as life became busy and I had children of my own, it became neglected and eventually took up residence at the back of a cupboard in the kitchen. My oldest daughter, who wants to be a chef, was looking through the kitchen cupboards one day when she came across the giant saucepan. Her immediate questions were: what do we need this for, and why are you keeping it when you never use it? Let me tell you a story, I said, about your *nanni* who made this pan with her own hands, and which was the greatest gift she could give her youngest daughter. A gift greater than gold.

As I told her the story, I could see that at ten years old she couldn't quite grasp what I was saying. I watched with understanding as she listened to this strange story about an old piece of cookware, took it in, and quietly filed it away in some dusty corner of her mind. One day, my daughter, you will understand why I have, in my small house, held onto this big old saucepan which I rarely use, even though it claims an entire cupboard. I hope one day she will one day take out of her own kitchen cupboard, dust off, and understand what mum was trying to tell her that day. Perhaps, one day, she'll even make jam with it.

I hope this story goes some way to demonstrate that this book isn't just about environmentalism or nuclear energy. It's about life itself, and the foundation upon which all of our stories are built. From the moment early humans first harnessed fire to cook meat, stay warm, fend off predators, and perhaps even spark the development of language, energy has shaped our evolution. It has powered every leap forward, from the growth of our brains to the rise of civilisation. Fire not only strengthened us by preserving our food – it shaped us as storytellers, drawing us together beneath the stars to share our earliest myths.

Those myths have passed through generations, contributing to a collective experience that shapes our understanding of the physical world. Cultural stories contain certainties, deceptions, half-truths, pseudo-truths, metaphors and pure fiction, but they influence social opinions, particularly in the field of scientific and technological innovation.

I will get to the myths and the pseudo-claims, but the question I want to ask with this book is: what could we achieve with true energy abundance and sharper understanding of the choices we are making? Energy is the engine of human progress – it lies at the heart of our past, powers our present, and will determine our future. Our decisions demand clarity and wisdom, thought and foresight.

Because what we choose to do next with energy will shape the fate of humanity.

PROLOGUE

In a different timeline, I'm crouched in the mud over a crude wood-burning stove, cooking *roti* for dinner. My hands are thick with callouses from years of tending flames and kneading dough, but I'm used to that, and besides, all the women here have them. Maybe I have children in this life, though not with the support I had for the two I delivered in England, both with medical help I wouldn't have access to here. More likely, I'd have had and lost many children in this timeline, but accepted the losses and kept going, because what other choice is there? Contraception is a luxury we don't have access to, and anyway, children bring us joy, and many hands make light work. Such is life in the village.

There is no electricity here.

Stay with me in this other timeline for a moment. Had I been born here, in rural India, life would look something like this: the heat is beating down relentlessly on my skin as I swat flies from my face and my food. They don't bother me, but the mosquitoes are annoying; they bite without mercy and ease off only at nightfall.

In the distance, a pack of dogs begins to howl. They're far away enough not to be of much concern, but some of the village women call their children back from the dirt road anyway. They remember that time Sunny was bitten by a wild dog and developed a fever and a headache that wouldn't go away. When he started vomiting, we knew it was his time. The nearest hospital is four hours away and, without transport, what chance did he have? And even if we made it there, how would we pay for his treatment? Hospitals here demand payment upfront. People often pass away on their doorsteps.

There is no doctor in the village.

Even as his father raced to the next village to find the closest thing we have to a doctor, we knew it was too late for Sunny. We prayed to the gods as we watched him die.

In this different timeline I do not know how to read or write. I bear children, and I lose children. I have many siblings, and have lost many of them as well. I suffer immense loss, but I know no different – such is life in the village. We call it *kismat* – fate. Preordained. God's will. We have to rationalise it that way, because we have little control over what happens to us or our kin.

In forty-degree heat, the men tend the rice paddies. We eat rice every day, sometimes twice a day if the rains have been kind. We trade it for flour, which we use to make *roti*, another daily staple. There isn't much protein, just lentils and chickpeas when we can get them. Some Punjabi families own a cow and make all kinds of luxuries from the milk, but not us. Still, we are lucky in our own way: we have shelter and have basic tools for farming. Not everyone can say that. A short walk outside the village is a stark reminder; where homeless children sit in the dirt begging for coins, and disabled children get by on makeshift crutches, or, as I've seen more than once, use crude wooden skateboard-like contraptions as legs. Wherever you look, you see what people don't have rather than what they do have, and although the former outweighs the latter, you learn to focus on the latter.

There is no other choice in the village.

Our lives depend on the rains. But they are less predictable now, and don't come as often. A world of extremes that grows more extreme. I don't know why that is, or why it is so unbearably hot now on summer days. *Maa* says it never used to get this hot; but no one here keeps records, so perhaps that's just nostalgia speaking. When the rains fail, we have to eat less. That's hard, because daily chores are already tiring and I am always hungry. I worry more for my hungry children, though. I don't want them to grow up stunted like their cousins. During bad spells, you can see it in their bones, in how slowly they learn, in how often they fall sick. It's not their fault their bodies don't have enough fuel to grow or fight. What did we do to deserve this? We believe that *Waheguru* must be punishing us for something we did in a past life. I try not to blame myself for that, because it's not in my control.

When I look at my children, I try not to count ribs that I shouldn't be able to see. I break *roti* into smaller pieces so it stretches further; add extra water to the *dal* so it seems like more. Sometimes I tell them I've already eaten so they don't feel guilty about my hunger. I don't know if they believe me.

There is no supermarket in the village.

Hunger becomes normal after a while, but it still leaves a mark. I worry that years from now, even if things get better, they'll still carry this emptiness inside them – physically, emotionally. I want more for them than just surviving. I pray to the almighty *Waheguru* to keep them safe.

In this timeline, I am crouching and coughing over the small outdoor charcoal stove on which we cook all our meals. I do this several times a day, to make tea as well as food. I wrap my *chunni* tighter around my mouth to reduce how much smoke I inhale, but still my throat burns and my eyes sting. In an hour, I'll be finished with preparing the food, and then we can eat. After that, I'll clean up the ash and gather wood to burn on the stove tomorrow. The clay pot I cook in is starting to crack and may need replacing soon, so I'll have to improvise with some old tin cans I've been saving from when we had visitors years ago. Family from abroad who ate baked beans with *roti*.

Everything is hot – my skin, my black head of hair that absorbs the heat. I pull my *chunni* up over it, and pretend that it is helping. Again, we are lucky to have this stove, and the pot, and a *tawa*, and not have to cook on an open fire like some of our cousins do. Cooking over a large fire in this heat is unbearable, but I bear it, because there is no other choice.

Heat is an all-consuming sensation: devouring my bare feet as they grasp the hard, dry mud, emanating from the stove in waves that distort the world behind it, licking my hands as I tend to the fire. In the real timeline, I would be outspoken about my suffering, and I would find a way to improve it. But here, I know no different, and I cannot change it, so I will not complain.

There are no microwaves in the village.

I wonder how Cousin Amit copes with this heat. He is confined to a wheelchair, although it's not really a wheelchair, because we are not in the real timeline. Bear with me. Amit is tied to a plastic chair with rope because he is disabled and can't sit unaided. We've done our best to reinforce the backrest and attach sticks to the legs so the chair can support his weight.

Amit was not born disabled. He suffered epileptic fits at birth, and a kindly doctor from a neighbouring village advised his parents that baby Amit needed urgent medical attention to stop the fits and prevent brain damage. It was not an uncommon problem, but he knew (as I did) that accessing the medication needed to stop the fits was also uncommon.

They tried to get hold of it for him. But it meant making a four-hour trek to the hospital, borrowing a motorbike, paying for petrol, and then paying upfront for medication that no one had spare money for, and having to do this regularly, because medicine can't just sit around in forty-degree heat.

There is no refrigeration in the village.

There is no overnight delivery in the village.

There are no wheelchairs in the village.

So Amit sits. Day and night, he sits. In his wheelchair-that-is-not-a-wheelchair, he stares straight ahead until someone moves the chair so he can stay in the shade or face a different direction. He drools, so we wipe his mouth for him, and we try to keep him hydrated. Sometimes he makes noises – sometimes cries out – which means he needs something, like help going to the bathroom, or water, but it's hard to always understand what he wants. Several times a day his *maa* and sisters lift and carry him, feed him, clean and wipe him.

I wonder if, in another timeline, I might have been Amit.

They care for him without complaint, but socially he is seen as a burden. Not just physically, but in the eyes of the villagers who believe that such a poor fate can only have come to pass as the result of his family's past wrongdoings. We call this *karma.* It's rarely spoken aloud in the village, but it hangs heavy in the air, felt in the mean glances of hungry neighbours when food is scarce, and Amit, spoon-fed, has rice smeared down his front. Disability is seen as a luxury here; a mouth to feed that doesn't contribute. Most families hide it as best they can, but Amit likes to be around people, and it wouldn't be fair to confine him to one room (we have one room; a simple shelter in which the whole family sleeps at night), so we bring him outdoors to sit while we do our chores.

If life is so cruel here, why am I imagining being born into this other timeline? It would be easier not to think of it at all, and I know you have never heard the story of this village before, because we are the people without a voice. But we also have a story that needs to be told.

My village isn't marked on any map; there is no GPS here. The signs are crudely painted on walls of crumbling buildings, written in the local dialect by those few who have come and gone who could write. Google Translate cannot help you here, or find your way to navigate these villages – which are homes to millions of people – you would need a local guide and some form of pidgin. But it doesn't matter that you can't find us because travellers do not visit, and people do not leave except by way of *Waheguru*.

We have our own stories, but you never hear them, because we have no way to communicate with the world beyond the village, and anyway, who wants to hear from the poorest people in the world? Poverty makes you sad and you'd rather look away from it. Besides, there is no easy way to help us. We cannot read or write, and we do not appear in documentaries. We appear in statistics: poverty rates, infant and maternal mortality rates, and other numbers you read in your newspapers over your morning cup of tea, which you brewed using an electric kettle.

To you, we are just numbers.

And there are millions of us.

And you blame us for that, too.

Thankfully, that is not my timeline. I did not grow up in the rural Punjab, because my parents were given the option to leave India, and they took it. They were given the rarest thing of all in that part of the world: a choice. They left behind everything they knew – family, culture, food, language. My grandfather fought for the British Army in World War Two and was held as a Prisoner of War in Japan. He later escaped – one of only a few thousand to do so – but I know little about this time as my dad doesn't like to talk about him. I have tried to elicit answers, but whenever I try, his eyes cloud over and he goes quiet from the painful memories. Granddad likely suffered from undiagnosed and untreated PTSD and was prone to fits of mania and violence. As recompense, and because Britain at the time needed workers, he was offered relocation, and when he arrived in England with his children he became a foundry worker. While the years of toil never ended for him, he never complained once he was here. Life was hard in the village. This is something my parents have also never forgotten.

Britain in the 1960s was full of opportunities for people who wanted to do manual labour, and my grandfather settled in Birmingham during a mini industrial boom when many factories needed many hands. Mum is vague on the exact date she arrived, but Dad tells me he arrived in England in 1963, at the age of twelve. They came by sea and changed boats many times, as (he says) there was no direct way to travel by plane then, or perhaps there was but it wasn't financially an option. They initially lived in Handsworth, Birmingham, in the middle of the country. As a child, my father did something generations of his people never had the chance to do: he went to school and learned to read and write. He was exceptionally bright, and as a child I remember watching him read his treasured medical encyclopaedia set, and how he would often diagnose us before a real doctor could. It was his dream to become a doctor, but he had no money for further education, and anyway, he came from a people that did not have the luxury of dreaming. As soon as he left school he went to work in factories, packing and lifting boxes and running machinery until the day he retired. He was never once late for work, and never took a sick day in over forty years at the same company. He didn't want to stop working, but when the factories shut down as industry moved abroad, he had no other choice. But, as we know from the alternate timeline, choice is relative.

So this is where the real story begins. Not on the hot mud floor of a nameless village in the Punjab, but in the heart of an industrialised nation. The first of my generation to escape manual labour, to have a dream, to attend university, and to have a voice; that's me, in the real timeline. And I'm using it to tell these untold stories.

CHAPTER ONE
A RUDE AWAKENING

'Nothing is so painful to the human mind as a great and sudden change.'
Mary Shelley, *Frankenstein*

'Change is the essential process of all existence.'
Frank Herbert, *Dune*

I was under bright studio lights and close to a million people were watching. It was like a classic anxiety dream, except, in this case, there was no chance of waking up.

Across from me in the BBC studio was the formidable Andrew Neil. The Scottish journalist and broadcaster was seated behind a plexiglass desk emblazoned with his name and initials. There were not many staff members around, although journalist Laura Kuenssberg nodded a curt hello as she walked past. The make-up woman asked me if I was nervous. 'He's friendlier than he seems,' she told me reassuringly.

Only four hours earlier, I had stood on a makeshift stage in Trafalgar Square, speaking about the imminent danger of the climate crisis, while around me police officers dismantled the tents of fellow protesters, dragging them away even as people and their belongings remained inside. In my role as a spokesperson for the global environmentalist organisation Extinction Rebellion (XR), the BBC had booked me to speak about the day's protest. I had XR's support. Before I left Trafalgar Square, one of the founders of the campaign group told me to close my eyes and stand against a tree as she performed a pagan ritual for luck.

A familiar face to UK viewers, from his appearances on news

shows for more than three decades, Neil had his own prime-time talk programme. He was widely admired for his no-nonsense way of getting to the essence of an issue, a style of interrogation he cultivated during a career at *The Economist, The Sunday Times* and other publications, in addition to scores of broadcasts. Though he is a known political conservative, Neil does not play favourites when putting his interviewees on the grill. A few months earlier a devastating exchange with the American right-wing political pundit, Ben Shapiro, went viral after Shapiro lost his temper under Neil's withering questions.

I had been briefed by the media team at XR that Neil was a climate change sceptic, so in my role as XR spokesperson I knew I was walking into the lion's den. I was a bit nervous after agreeing to appear on *The Andrew Neil Show*, but in the familiar phrase voiced by many a seasoned veteran, this wasn't my first rodeo.

I felt that this interview promised a valuable opportunity to talk about climate science and explain our protest to a large national viewership. The show's announcement on social media in the hours before the broadcast had caused me to do a double take: I was tagged as the guest to follow former prime minister, Tony Blair.

That afternoon, while walking across London towards the BBC studio, my phone pinged with messages from people wishing me luck and telling me which talking points to push. They sent facts and figures, but I already had them memorised. A text message from a BBC Asian Network journalist, who had previously worked with Andrew Neil, advised me to speak up. 'He appreciates people who stand up for themselves,' she advised me.

After walking past pro- and anti-Brexit demonstrations outside Westminster, the empty BBC café served as a quiet place where I could collect my thoughts and let the London street noise subside. I lay on the sofa and began concentrating on the task at hand. I was here to communicate the dangers of climate change and call for effective climate action. I gave my notes a quick glance and double-checked the facts I might need to quote at a moment's notice. I felt confident and prepared.

I was wearing a necklace I had made when using polymer clay with my daughters. It was bright green – one of the XR branding colours – and it was shaped into a crude hourglass – our emblem. XR's message to the world was: 'We are running out of time.' My message to myself was: 'You are doing this for your children, and for future generations, so you'd better get it right.' But mostly, I wanted my daughters to see something they had made on television.

The studio set was the least friendly I had ever encountered. This was no casual chat show set with sofas and end tables. I sat on a small stool under a spotlight some distance away from Andrew Neil, who was wearing an earpiece and installed behind his TV studio power desk. In front of him was a sheet of paper with questions and notes on it.

After a brief introduction Neil checked his notes and began the conversation with, 'Some of your activists claim on TV that billions of people are going to die in quite short order.'

My heart sank. I had been briefed by the XR team on what the interview was expected to cover. Among other topics, I was prepared to discuss our latest attempt to raise awareness by blockading key roads around Westminster, the intricacies of climate science, and the idea of individual versus collective action. Clearly, Neil had a different approach in mind.

'One of your founders,' he said, referring to his notes, 'said in April, "Our children are going to die in the next ten or twenty years".'

Whether by pure luck or the result of excellent research, *The Andrew Neil Show* staff knocked me off guard with their opening question. It was one I knew I could not defend.

I hedged. I explained that climate change would lead to massive population shifts, and tried to avoid addressing specifics of the forecast.

Then came the logical follow-ups: 'What's the scientific basis for these claims?' and, 'Aren't you scaring people with this rhetoric?'

While over a million viewers watched my visible discomfort during the broadcast, part of my mind was telling me that all my years of work were spiralling down the drain in a few short minutes – the Masters in Science Communication that I had studied part time while working and raising two children alone; the years of writing articles and public speaking, debunking myths and combating pseudoscience... it all crumbled before me.

Rob Burley, one of the producers of the show, later wrote in his book that this was his 'favourite interview'. But, at the time, I felt as though I was watching my career go down the drain, and I questioned everything I had done in the environmental movement up to this point. I felt overcome with a feeling of failure.

For a long time I had sensed climate change to be the biggest existential threat of our time. The science left no doubt in my mind, and I had dedicated years of my life trying to effect change before it was too late, but I could not defend the forecast of billions of deaths in the next two decades. This was a story made up by another XR spokesperson and defended by many XR members. I knew it was wrong and unhelpful, and I could not bring myself to defend it.

Viewers were unaware, as they watched me freeze like a rabbit in the headlights, that I had already taken umbrage with this figure of six billion deaths when it was first stated publicly by one of the co-founders of XR some months earlier. The co-founder had assured us not to worry about it, as he had made a 'reliable back-of-the-envelope calculation' regarding the number of deaths. I had pushed back, arguing that it was unacceptable to make up numbers while also expecting people to believe the science. The debate had raged for weeks, and my side ultimately lost; the XR spokesperson refused to retract the figure, and XR's Media and Messaging team backed him up. The smallest concession I had obtained was that, given the negative press following this statement, similar comments should not be made by spokespeople in future. I was also told never to comment on the statement publicly.

So that put me in a difficult position.

Of course, none of that was public knowledge. As I sat there under the spotlight, questioning how I had ended up in this position, I knew that I was meant to avoid the subject, I personally felt that I could not defend the figure, and also knew that thousands of my peers were watching this broadcast live in their homes and at the XR office, and they were not cheering me on.

Sitting in that BBC studio, it became blatantly obvious to me that what I had feared earlier – using misleading facts in an attempt to sway public opinion – is ultimately counter-productive. All too often, it leads to disillusionment, cynicism, distrust and apathy. Influencing public attitudes to effect change relies on the presentation of clear, verifiable facts, and doing so additionally requires dealing with the messy realities and ambiguities that sometimes surround these facts.

In the case of the environmental movement and climate change, it means confronting an emotionally fraught alternative that no one wants to discuss.

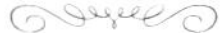

The fallout from the interview was immediate. When a clip appeared on Twitter the next day, it rapidly amassed two million views. People took to commenting that I exemplified everything that was wrong with the climate movement. I was called clueless and stupid. The tabloid press hated me for my inability to retract a statement I had not personally made or stood by; members of XR were angry with me for refusing to stand by the statement or successfully change the subject to something else; and everyone seemed annoyed with me for wearing a neon green necklace of the hourglass symbol with such confidence.

The interview was all over the news, and the takeaway left my credibility in tatters. I switched off my phone and ignored social media for a while, but this didn't stop XR supporters and members from contacting me to tell me how poorly I had performed in the interview. They told me I should have defended the figure; I should have pivoted to talk about citizens' assemblies; I should have been better prepared.

As the number of news articles calling me clueless ramped up, and were used to deny all climate science, I realised I had to go public about why I had not defended the figure. Andrew Neil's point about the alarmist rhetoric around 'billions of deaths' brought to the surface an argument about public messaging that had raged within XR for weeks. Back when the statement predicting billions of children's deaths was uttered on television, several other scientists and I had argued heavily for retracting it. But our entreaties were not heeded, and this was a direct consequence of that. To me, this seemed like a deep failing, not just of XR, but of the environmental movement in general.

Ultimately, I felt let down, and I had realised in that BBC studio that I could not tell this movement's story any more. Andrew Neil's probing questions requesting solutions for cutting zero-carbon emissions before the end of the decade had also put me in a bind. As a spokesperson for XR, I was unable to offer him solutions and, exasperated in the moment, I had told him so on live television.

Officially, XR's position is a demand to Go Beyond Politics; a call for a new political system to deal with climate change. By linking climate change to radical political reform, XR had persuaded much of the climate activist movement, including Greta Thunberg, that this was a realistic path forward. Personally, I regarded such a dramatic political transformation as unrealistic, and disproportionately time-consuming when the need to tackle climate change was already so pressing, and I had questioned it openly within XR many times. Needless to say, I had not changed many minds there.

Neil could tell that something was amiss, and he continued to press me on topics I could not answer. Whoever was guiding him in his earpiece knew what they were doing. He asked about gas use: 'How much would we need to give up?' 'How are we meant to live without it?'

Of course, the answer XR would have given, though not an official policy, would have been to build more renewables, to reduce energy usage, and to remove fossil fuels at any cost. But this position did not reflect the scientific consensus, and I knew it. I might have looked calm in my chair, but an inner battle was raging. I wanted to express what I had come to realise: that the scientific consensus is that we should build nuclear power plants *as well as* renewables in order to decarbonise. This is what the Intergovernmental Panel on Climate Change (IPCC) has demonstrated; all of the decarbonisation pathways by Working Group III, in the landmark *1.5 Warming Report*, require significant development of nuclear energy. Yet this was something that most of my colleagues in the fight against climate change had long regarded as heretical. I was in a no-win situation, and I knew it.

Since 2015, when I authored a book on solutions for addressing climate change, my thinking has evolved. Sitting under the lights representing the world's most prominent non-violent climate change activist organisation at the time, I was constrained from expressing what I had become convinced was the realistic answer to Neil's question about gas.

I knew I could not mention nuclear energy in a positive way in my role as a spokesperson for XR. Firstly, I felt a duty to speak only as a representative of the group, since that was what I had agreed to do in this interview. But secondly, four years earlier I had included a chapter endorsing childhood vaccination in *The Ultimate Guide to Green Parenting*. As a result I received hate mail from the anti-vaccination movement and was repeatedly removed from the roster of speakers at green parenting forums. I had been here before; swum against the tide, angered my community, and lost friends in the process. So, in the moment, I decided to stay quiet.

Were I to even raise the possibility of nuclear energy as a solution to climate change, I knew there would be a massive backlash. Sadly, back in 2019, it wasn't even a subject for discussion. I had protested against nuclear energy for years in my early twenties. There had always been a pervasive anti-nuclear sentiment within the environmental movement. Sentiment is easily absorbed in solidarity with like-minded people sharing a common goal – you simply accept it to be true. As part of the group you are not encouraged to question the dominant ideology on an issue, nor research and present your own ideas. So, for many years, I didn't, and I would have reacted negatively if someone had tried to change my mind. The irony, in hindsight, is that even as I was arguing with anti-vaxxers, explaining how they had the science wrong, I myself had the science wrong on nuclear energy.

How had this happened? Good stories enter the realm of myth among people sharing their sentiment, but only when we step outside the group, can individual perspective and fresh truth enter our minds and establish a foothold. Stepping out is an immense risk to all that you hold dear, and it is often easier to stay quiet than to ruffle feathers. But since, in the present situation, feathers had already been irrevocably ruffled, I realised it was time to step up.

I had already been in favour of nuclear energy for many years before becoming involved with XR, but I had never spoken about it publicly. For most of my life, the environmental community had told me one story – that nuclear energy was bad – but in recent years, science had told me a different story, which hinged on the uncomfortable fact that I had been lied to. That is a difficult thing to accept, and it hit my identity hard whenever I thought about it.

Since I had believed the stories told by my community about nuclear energy, throughout my twenties I was actively anti-nuclear, and the process of changing my mind was not an overnight venture. For many years I was a member of the UK Green Party, and had even been asked to run as a councillor by my local group. Like the traditional Greens, I believed that nuclear energy was bad, while solar and wind power were the sole solutions to combatting climate change. I had championed the idea of transitioning the electricity grid to 100% renewables. But as I spoke to climate scientists about the solutions, read their reports, and studied for an MSc that further increased my ability to decipher and question, I came to accept that we also need a back-up energy source to supplement renewable energy on those days when the sun and wind are lacking. The need for back-up energy was never questioned by my peers, as the Green movement held to the idea that infinite battery storage would one day solve the problem and was 'just around the corner'. It was a phrase I heard them use often and had used myself when lobbying my university to switch to a renewable energy provider, which was a campaign I helped to win.

At the time, for anyone who was ideologically eco-conscious, nuclear energy and fossil fuels were already off the table. The dominant belief was that using energy had caused climate change, therefore energy usage was bad and we had to wean ourselves off it, meaning that building more plentiful sources of energy generation was a bad idea. Wind and solar power fit in nicely with this ideology, as their intermittent nature meant that we would never have an abundance of energy from them. While these are not opinions that most people associate with green campaigners, they absolutely underpin the causes they choose to fight. And, beneath all that, lay a subtle but foundational ideology that I didn't understand for a long time – the notion that people are bad, because humans cause pollution and harm the environment, so *we don't deserve* to have access to abundant energy.

Since I was oblivious to the deeper ideology, I hoped that we were right about battery storage, as I did not believe that energy scarcity would solve the world's problems. Instead, decades later, and following significant financial investment in the technology, battery storage has not advanced enough to reliably support civilisation. It turns out that the technology we were awaiting for decades was *not* 'just around the corner'.

Current battery technologies, such as lithium-ion, face limitations in energy density, cost and scalability, leaving it a challenge to store enough electricity for periods when the sun isn't shining or the wind isn't blowing. That's where back-up or baseload power comes in, and historically it is always provided by either fossil fuels or nuclear energy. It sounds very simple when put this way, but for me accepting this new story completely rocked my world.

It's always difficult to stop believing in stories that you grew up with, especially when all your friends still put their faith in them. It's like learning that the Tooth Fairy isn't real, or Santa was a lie; there is a deeply felt sense of loss and magic, and perhaps even a sense of betrayal. It is far easier to cling to the old story. I was well past this stage by the time I did the Andrew Neil interview, but it still made me uncomfortable to think about it. The idea of admitting on live television, to millions of viewers, that all my friends still believed in Santa and the Tooth Fairy, felt like a step too far.

In truth, I had hoped never to clash with my tribe on this subject, but, as the saying goes, 'truth will out', and eventually having to go along with a story that you know is false becomes unconscionable, even painful. As soon as you realise that the story is built on weak foundations, you are at the mercy of glaring truth rearing its head time and time again, demanding that you look at it – pretending you can't see it becomes tiring.

Although none of my friends knew that such a shift was taking place within me, I started to decline invitations to protests, and as I shifted my perspective of what wind and solar power were capable of, I began to identify other aspects of the stories that didn't quite fit with the wider narrative. For example, in the many environmental groups I was involved with, minimal attention was paid to the environmental and resource impacts of solar panels, wind turbines or battery production. They were seen as 'clean and free' sources of energy – as if they require no resources, mining and manufacture. Raising issues like this, however lightly, had landed me in trouble with my community before, so I learned to avoid the topic altogether. Whenever someone told me that governments weren't building wind farms 'because it's free energy and can't be controlled by the man', I would bite my tongue. But after a while, my tongue began to hurt.

More crucially, my fellow environmentalists would argue that all energy usage is bad, but at least wind and solar power would forcibly limit how much we could use – 'no one needs to watch

TV at night' was a common saying in these circles. This position clashed with my life experience – my mother and her saucepan – and opened a can of worms that I didn't feel able to untangle. I set it aside instead.

Eventually, I saw through the façade of the Tooth Fairy and knew it was time to thank my parents, not some magical nighttime being, for the money left under my pillow in exchange for a tooth. After many years of hearing the same arguments and watching as the same questions were avoided, I became more forthright in asking why were we protesting nuclear energy and fossil fuels, but had accepted 'renewable' energy as the perfect technological solution?

The main argument I made appealed to views my peers and I still had in common: that renewable energy is not created equally.

While biomass is technically classed as 'renewable', it can be heavily polluting depending on how it is used. Burning wood, crop waste or dung, especially in open fires or inefficient stoves, releases harmful pollutants like fine particulate matter and carbon monoxide, contributing to respiratory illness and premature deaths. On a larger scale, biomass power can emit more carbon dioxide than coal if forests are cut down and not replanted. While using waste materials and modern technologies can reduce its impact, biomass is not inherently clean and should be managed carefully to avoid environmental and health harms.

Also, biomass confuses our statistics for emissions by forming the bulk of figures for renewables around the world. For example, in the UK, biomass is the biggest source of 'renewable' energy consumed, and the main source of biomass burned in UK power stations is wood pellets. While it looks good that countries are investing more in renewables, the reality is that we are burning trees to do so, which defies the objective, considering what we are attempting to achieve in decarbonisation and net zero targets.

Scientists have drawn attention to this in recent years, and a letter signed by 650 scientists[3] argues against the use of biomass:

> 'Troublingly, because it has wrongly been deemed "carbon neutral", many countries are increasingly relying on forest biomass to meet net zero goals. This is harming our world's forests when we need them most. Many of the wood pellets burned at power stations for bioenergy are coming from whole trees – not wastes and residues from logging, as the industry claims. For example, nearly half of all biomass burned at the UK's Drax Power Station comes from whole trees.'

Again, that's not to say that biomass doesn't have a role to play at all, but we need to ask why it has a more positive image than nuclear energy.

Then there's hydropower. While hydropower is clean and more reliable than wind and solar power, it can cause immense harm environmentally and to local populations wherever dams are built. It can also be very dangerous: the 1975 Banqiao Dam[4] failure was not only the collapse of the Banqiao Dam under the influence of Typhoon Nina, but also sixty-one other dams in Henan, China. 26,000 people died in the floods, and an estimated 145,000 later died from epidemics caused by water contamination and from famine. Estimates put the total death toll at more than 220,000 people, with over ten million people affected by the disaster.

Take that in for a moment: that's more deaths from hydropower than from all the combined nuclear meltdowns in history, including the worst three meltdowns that we are familiar with. Yet people do not protest hydropower and politicians do not ban it. Nor am I arguing that we should do so, but it's interesting how the story of hydropower is positive for most people, and doesn't conjure up images of destruction in the way nuclear disasters do. As environmentalists, we never protested or worried about hydropower, which is immensely more deadly than nuclear energy, and causes significant harm and displacement when it goes wrong.

The only way to overcome the stories that have shaped our ideas of these technologies is to start from scratch: to identify where those stories came from and why they were told, and consider the bigger picture when discussing energy sources. Only then can we see the wood for the trees.

Why do traditional environmentalists want the world to be run on 100% renewables? This relates to a bigger story, about 'The Man'. In the hippy circles where I spent my younger years, all of humankind's problems were blamed on capitalism and wealth. This was referred to as 'the system', or The Man. An argument we made throughout the '90s and early 2000s was that The Man didn't want us to have 'free' clean energy like wind and solar power because if we stopped burning fossil fuels then rich people would lose money. The logical conclusion of this was that no one would invest in renewables because it would make us all rich and independent. I used to believe this story, which I heard countless times in numerous ways, not just at activist events but also in popular books and films.

So when I first heard that Germany was attempting to transition to using 100% renewables, I thought the tide had turned. A friend said, 'We're winning! Hooray for the Germans!' While, privately, I didn't agree with their stance against nuclear energy, I thought it didn't matter whether we used nuclear or not, since we were about to see evidence for how to transition to 100% renewables without the need for fossil fuels or nuclear. We were about to gain evidence for something we had known to be true for decades. It was about time.

In the 1990s, over a third of electricity in the newly reunited Germany came from seventeen nuclear reactors, but Germany decided to phase out nuclear energy in the early 2000s, as part of its *Energiewende* approach to rely only on renewable energy. Initially the decision was reversed, but then in 2011, when the disaster in Fukushima happened, German politicians moved forward with the plan to entirely phase out nuclear energy, and this decision was largely backed by the German public.

This became the pivotal moment when the last pillar of the story of free and abundant renewable energy began to crumble. I watched with disbelief as Germany struggled to fulfil its mission, and in an effort to replace lost nuclear-generation capacity, Chancellor Angela Merkel signed a deal with Russian President, Vladimir Putin, to build the Nord Stream gas pipeline. Although this was years before Russia's full-scale invasion of Ukraine, I already felt uneasy about increasing energy dependence on an authoritarian regime. It also felt like a slap in the face, considering that 100% renewables was supposed to mean 0% fossil fuels.

Meanwhile, the financial impact of Germany's transition to renewables was far greater than anyone had predicted. Energy prices began to rise sharply to fund the solar and wind buildout, a burden I felt was unfairly passed on to ordinary citizens. Still, this was deemed necessary as part of the energy transition and was treated as a temporary blip. But then, maintaining grid stability became increasingly difficult, and by 2021, the German government had spent billions of euros[5] managing transmission congestion, including paying conventional power plants to reduce output when the grid couldn't absorb the excess. Then Nord Stream was sabotaged and the energy crisis deepened. I watched in disbelief as climate activists protested against massive mining machinery to stop an entire village from being bulldozed to expand lignite coal mining. This was already years after Germany was supposed to have transitioned fully to renewables, following its planned nuclear phase-out.

The initial transition saw a rise in coal usage and reliance on imports to compensate for the loss of nuclear energy. This reliance on coal resulted in higher energy prices and potentially negative health impacts from increased pollution. The shift created an increase in carbon dioxide emissions, with one study estimating an additional 36 million tons per year and associated health impacts from respiratory and cardiovascular illnesses. The truth was impossible to ignore: the vision of *Energiewende* had failed to materialise as promised. I could no longer accept the story that The Man was standing in the way of renewable energy – it was evident that it had its own drawbacks. Wind and solar power alone could not power a large, industrialised nation, let alone the world.

For me, the conspiracy of The Man blocking progress on renewables fell apart, but, to my surprise, none of my peers felt this way. Instead, they made excuses: no one could have known what Putin would do; it doesn't matter if they need to burn coal for now; at least they helped make solar panels cheaper for everyone else; the plan is still on track; who cares what it costs. And so on.

I felt conflicted, like a child who has seen behind Santa's veil but whose siblings still believe the fairytale. Only, in this case, believing in the lie meant losing out on presents. I worried that we were championing the wrong solutions – worse, that I had been backing the wrong approach for most of my life.

What if we were at fault for fighting for an ideology that had not only failed, but was exacerbating climate change and harming the environment further? Wouldn't it have been better to at least keep the nuclear plants running while trying to set up a brand new wind and solar power grid? And wasn't burning more coal the opposite to our goals?

How could my friends defend this?

Again, my working-class background set me apart, because I also worried about the cost of this energy transition to ordinary people. The total cost of *Energiewende* has now been calculated at over one trillion euros. As well as contributing to government debt, it hit consumers hard – both families and businesses – for whom electricity prices rose sharply as *Energiewende* expanded. Between 2005 and 2014, residential electricity prices in Germany increased by more than the average total residential cost in the US. Businesses shut down, and people struggled to stay warm in the winter. In early 2020, electricity prices for household consumers in Germany were the highest in Europe.[6]

While renewables-only proponents continued to argue that increased CO_2 and dependence on lignite coal were necessary for the energy transition to succeed in Germany, I turned to scientific literature to find answers and solutions. To my surprise, the case study that had already proven how to reduce fossil fuel dependence and, in turn, lower emissions and air pollution, came from France.

As my belief in an energy transition based solely on wind and solar power began to crumble, my fellow environmentalists ignored France's case study, a proven path to success that took roughly a decade, while Germany's equally long transition has yet to achieve its goal of relying entirely on wind and solar.

Energy crises are not new; the oil crisis of 1973–74 caused the worst recession in Western societies since World War II. France and Germany were heavily impacted, but the countries took different approaches. German support for nuclear energy was very strong following the 1974 oil price shock, but we the public and the government later turned against nuclear energy.

In France, a different approach was taken. A common saying

at the time was, 'We don't have oil, but we do have ideas', and the country responded to the crisis with the Messmer Plan to go 'all nuclear, all electric', named after Prime Minister Pierre Messmer. France achieved energy security in roughly twelve years through extensive use of nuclear energy, and at its peak, close to 80% of French electrical consumption was nuclear.

France maintains a primarily clean and reliable energy grid to this day, yet their story was not common knowledge globally for some time. I stumbled on the facts while reading up on historical energy transitions. Science had already told this story, but no one else seemed to know about it, and when I mentioned it to an activist friend, she replied that it was false – 'made up by the nuclear industry'. Her comment helps explain why, when I first began my nuclear advocacy, France's success story hadn't yet reached mainstream attention – it didn't fit the narrative believed by the loudest voices for climate solutions and the energy transition.

It was difficult to come to terms with, but eventually I realised that for many years I had believed and been retelling the wrong story. I had supported gambling on an expensive experiment that only prolonged dependence on fossil fuels, when, all along, a neighbouring country showed us how a clean and reliable energy source could power our lifestyles without damaging the planet.

By the time the *Andrew Neil Show* interview was over, I realised that knowing this was not enough, and failing to tell this story sooner was where I had really erred. I now felt strongly that someone had to start telling the truth widely, as not a single person in the UK was doing so successfully. It was time to take up the mantle.

In my early twenties, I asked a question about nuclear energy at a conference hosted by the local chapter of the Green Party, of which I was a member at the time. I was not exactly 'pro-nuclear' then, but curious about why there was such vehement opposition to the technology in the environmental movement. I was in the 'questioning' stage of changing my mind.

After the discussion about climate solutions had ended, the audience was invited to ask questions, and I was selected from the crowd. I directed my question at the climate scientist from the Met Office who was on the panel, and asked him whether it was possible to tackle greenhouse gas emissions without nuclear energy. The host of the event immediately shut down my question, which concerned me, as it was not unlike my experience of speaking to anti-vaccination activists who did not allow free debate. Instead of handing the microphone to the scientist, the leader of the local Green Party responded, 'We don't need nuclear energy!' The two hundred or so people in the audience applauded her response, but for me, alarms bells were ringing. Surely, if this was true, the climate scientist would have been permitted to answer my question.

After the event ended, I sought him out and posed the question again. He replied, notably quietly, that the scientific consensus is that *there is no doubt* that we need nuclear for the energy transition and to reduce greenhouse gas emissions. My world spun. Not only had hundreds of people been prevented from hearing this difficult truth, but it had almost been kept from me too. Had everything I'd been told by my peers and friends been a lie?

On my way out of the room, the leader of the Party walked over to me, placed her hand on my shoulder, and asked with genuine concern, 'You're not pro-nuclear are you?' I mumbled that I was open to considering all solutions to fighting climate change. But her manner highlighted to me that I did not belong with these people, and would never reach common ground with them.

I ended my membership with the Green Party the same day,

writing to them that I could not be part of a group that did not allow free discussion of evidence-based climate solutions. Sadly, speaking out meant losing my community and friends too – a consistent theme in the story of my life.

When I told a friend what had happened, an engineer by training, he sent me a scientific paper on radiation, which addressed my main concern about nuclear energy at the time. He said I might be radiophobic and thought the paper would help me. What caught my attention was the rational, calm way in which he, John, approached the topic. I had just left a group where asking questions led to being raged at or dismissed, so I was ready to listen to a different viewpoint.

John was a trustworthy and honest person who was not involved in the nuclear industry – which I viewed negatively at the time – and who had in the past defended me when anti-vaccine activists had sent me hate mail over my first book. He was openly pro-nuclear, and I told him I was open to hearing and considering his position. Finally, this was true.

Frankly, the paper he sent me blew my mind. Nearly everything I had been led to believe about radiation, especially relating to nuclear energy, was wrong. While it didn't immediately change my beliefs about the technology, it stayed on my mind. I felt embarrassed, as I realised I had inadvertently been spreading misinformation – something I had spent years calling out with anti-vaccinators and climate deniers. I started questioning my other beliefs about energy production and usage, ones also acquired from the dominant environmental ideology who, as I now saw plainly, did not allow open discussion of this subject. Worse, because they harboured their own fear and were not willing to look at the reasons for this fear, they intentionally spread fear in others as well, which prevented people from assessing the facts rationally or honestly.

Still, it is a difficult thing to change your mind. Challenging someone's ingrained beliefs is naturally upsetting, because it threatens their sense of identity, and as a result of that people often take even the simplest of questioning as a form of personal attack. I experienced this in the following months, as when John tried to follow-up the research paper about nuclear radiation with another paper on the outcomes of Fukushima, I felt defensive and told him he was being pushy. My worldview was being turned on its head, and I needed time to process it. Why had I been lied to by my peers for over a decade? Who was I without my community, with whom I'd been arrested at protests; with whom I'd baked and broken bread? And what did it mean to be an environmentalist if you were opposed to climate solutions that had been proven to work?

As the years passed, I asked myself some unsettling questions: *It's been ten years, Zion, and you're making the same argument that you were making a decade ago… that renewables will save the world. Where are those long-term batteries to back up wind and solar that you were promised? And why are you still advocating to reduce energy consumption when, despite years of campaigning to reduce energy usage, the world is using more energy, not less? The best behavioural scientists in the world haven't been able to change this. This isn't about The Man, it's about human nature, and science… The best will in the world can't change the laws of physics. Look at Germany…* And so on.

I questioned everything, from the basis of all of my beliefs, how they had formed and who had helped to shape them. Somewhat naively, I initially attempted to share my new ideas with friends in other campaign groups, but of course they reacted badly when I questioned the dominant ideology. I suddenly felt very alone in the world.

Above all, I worried that, for over fifteen years, my actions as an environmental activist may actually have done more harm than good to the state of the planet. When the editor of a magazine on permaculture, for which I had written multiple articles for free, learned my views on vaccination, she emailed me to say that I was no longer welcome in her community. I knew that nuclear was a much bigger deal, so I stayed quiet, because I was tired of being shut down by people I cared about, and I was clearly a lone voice. I also wasn't ready to completely lose the little community I had left.

It was years later when, as the Andrew Neil interview ended, I recognised that I could not do this any longer, and more, that I didn't owe those people anything. I had kept quiet for too long and, frankly, it had only made me miserable. I didn't fit in with these groups. Perhaps I didn't fit anywhere. It was time to challenge unscientific ideology loudly and boldly. It was time to wake up and speak out.

Peter Parker's life was transformed the moment he was bitten by a radioactive arachnid. Amazing superpowers. Spidey sense. An awakened awareness of responsibilities to humankind. And a snazzy skintight red-and-blue costume to boot.

My own transformation into an environmental pro-nuclear advocate was hardly as dramatic as Spiderman's and it didn't happen because of a single fateful moment, although the BBC interview was pivotal in its own way. I may not be a superhero, but I confess to sharing Peter Parker's intrinsic fascination with science, his frustrated desire to make the world a better place, and his sense of bewilderment when others misinterpret his motivations. With great power comes great responsibility; and great culpability.

During my formative years as an environmental activist, I was probably one of the last people anyone assumed would become a passionate champion of nuclear energy, let alone promote it as a major technological alternative for reversing our global climate crisis.

For most of my life, I was the perfect environmentalist. I wrote a book on how to live with a low-carbon footprint, which won accolades across the movement and in the media, the front cover adorned with endorsements from environmental champion Bill McKibben and the then Green Party UK leader, Natalie Bennett. I signed a pledge not to fly and hadn't stepped on a plane in fifteen years. I went vegan in the early 2000s before most cafés offered non-dairy milk; I remember trying to melt the first vegan cheese ever produced and watching it catch fire under the grill. I co-founded the first environmental group on my university campus (the Campaigns Forum) and spearheaded a campaign to add fair trade and vegan options to the university campus cafeteria and store. I didn't own a car or learn to drive but walked, cycled or took the train everywhere. I lived my life as a protest. I organised rallies and protests, petitions and direct actions. I wasn't just 'doing my bit', I was doing *every* bit that was expected of me, or that I expected of myself, as someone who understood the serious risks of climate change. Back then, the narrative was that this would be enough to make a difference; that if everyone did it, individual action and self-sacrifice would save the world. Obviously, that turned out not to be true.

In my twenties, when Greta Thunberg was a baby, I had already been arrested multiple times for taking part in direct actions with a small protest group known as Camp for Climate Action. I risked jail when protesting the Royal Bank of Scotland's role financing the extraction of oil from tar sands, a process that both removed Indigenous peoples from their land and left the environment in ruin. We dressed in bin bags as 'tar sand monsters', slipped past security at an RBS branch in Edinburgh, and started covering ourselves in molasses to simulate tar sands. As security rushed at us, we sat on the floor to form a blockade and, as expected, they tried to grab me first. Due to my small stature I was always physically the weakest link.

In my mid-twenties I was risking everything because my concern for the planet was so great. What would I do if I was given a criminal record? How would I defend these actions before a judge? I'd already had these discussions with my affinity group, which is the name given to a small group that takes direct action in this manner. We all felt the self-sacrifice was worth it, because we were standing up for Indigenous rights and against significant destruction to the environment. As I walked into the bank and began to undress (the bin bags were hidden underneath our clothing) and pour molasses (to symbolise tar sands) over myself, I saw the security guards move towards us and reminded myself that I was doing this in order to help give a voice to the voiceless.

Around the same time as parading as a tar sands monster and being subsequently written up and banned from all RBS banks, I took part in an anti-nuclear demonstration as part of a human blockade that resulted in multiple arrests.

The night before heading south to Plymouth, my friend and I worked diligently into the early hours cutting out twin cardboard signs in the shape of gas masks. It was a cold November morning in 2010. I checked the weather forecast early and was of two minds about wearing my fair-trade poncho. It wasn't terribly waterproof and the possibility of rain was forecast. At Victoria Station we met up with other friends to catch the early morning train to the south coast.

My first anti-nuclear action wasn't protesting a power plant. Rather we gathered on that overcast November morning to fight the British government's decision to continue to upgrade and deploy its small fleet of nuclear submarines. To us, nuclear weapons and nuclear energy were basically the same thing.

It also seemed obvious that if nuclear weapons were produced and deployed on military vessels there was a likelihood they would be used at some point, either as part of hostile action or as the result of a careless mishap. Already, in the span of living memory, nuclear weapons had twice been dropped on civilians. And ever since 1945 the majority of the world's population had been forced to accept the existence of nuclear weapons as part of everyday life. The threat that, somehow, nuclear annihilation might arise at any moment was never that far away. There had already been a number of accidents and near disasters, many only coming to light years after they took place. Doomsday scenarios triggered by a careless mistake, or weaknesses in the established command and control procedures, had inspired screenwriters to produce nuclear age horror stories that became a staple of Cold War entertainment.

As part of my protest regalia I wore earrings displaying the Campaign for Nuclear Disarmament peace symbol, the ubiquitous Sixties' icon first seen in Britain during Cold War 'Ban the Bomb' demonstrations.

A half-century later, here we were continuing the protest; but countering the public's seemingly complacent acceptance of high-tech militarism was only one aspect of our nuclear submarine protest. The demonstration leaders at Plymouth's Devonport shipyard proclaimed that the facility itself was a real and present danger. Protest literature said the Devonport shipyard's very existence put the lives of the 250,000 people living nearby at risk because, 'the dockyard walls, while formidable, do not stop radiation'.

Militarism and the threat of nuclear radiation were further signs that so much was heading in the wrong direction in the new century. Increasing wealth inequality. Blighted cities.

The Plymouth protest would blockade the Devonport

naval dockyard, where the UK's four ageing Trident nuclear submarines were being individually refitted. There had been some talk that HMNB Devonport might become the fully-armed subs' new home port.

The protest literature dubbed it 'the Sellafield of the Southwest', an allusion to the notorious multi-use nuclear site on the North Sea Coast in Cumbria, infamous as the site of one of the world's worst nuclear accidents in 1957. Sellafield continued to have a sinister and frightening reputation. A book, published by an award-winning American author some years earlier, contended that decades of human death and disease had resulted from dumping tons of nuclear waste in the coastal waters, also destroying the natural environment. Sellafield became even more famous when *Kraftwerk* intoned its name – along with Hiroshima and Chornobyl[7] – in an electro-pop hit that reminded everyone that 'radioactivity is in the air for you and me… discovered by Madame Curie'.

In densely-populated Plymouth the natural world appeared once again under assault, this time under the guise of national security and defence of the realm. The Devonport shipyard was the battleground and it seemed time to take a stand against the twin threats of atomic weapons along with the diseases and death that would result from another careless Sellafield-like dump of nuclear pollution.

Many of my friends, non-violent activists, had been protesting the presence of Britain's four Trident nuclear submarines for more than a decade, and we were continuing their legacy. The four individual Trident subs were armed with forty-eight nuclear warheads, each with eight times the destructive power of the Hiroshima atomic bomb.

Our spirits were good, fortified by an awareness that we were doing something important. If our actions were successful we would assure a more peaceful and healthier future for everyone. It was sometimes difficult to understand why everyone wasn't marching with us that day.

Upon arriving we realised we were the youngest people at the protest. We joined what was a sombre march, holding high our cardboard gas-mask signs. In bold letters we had painted: NO RADIOACTIVITY! NOT FOR YOU – NOT FOR ME!

The protesters surrounding us appeared to be veterans of scores of earlier demonstrations. Some had possibly been marching regularly since the Sixties. Unlike most protests I had been part of in recent years, this was anything but a festive occasion. When our group began to sing protest songs, other marchers asked us to remain quiet to reflect the deadly nature of what we were protesting. Their sombre mood was in harmony with the grey and drizzly weather. I wondered how they had kept up their activism so long without a little music to balance out the bleakness.

For four and a half hours we blockaded the main gate, with some activists locking themselves into their own vehicles. One protester attempted to lock himself to a local police car. In all there were fourteen arrests, but I was not one of them on this occasion.

Among workers at the Devonport shipyard, and those in the defence industry who relied on the Trident submarine for their jobs, there's little doubt we were regarded as troublemakers – if not worse. But being arrested is easily rationalised as a badge of honour when convinced that you are in the right, and feel that posterity will eventually recognise that you were on the right side of history to protest militarisation, pollution and deadly environmental degradation.

We had no doubt that everything to do with nuclear technology was bad. Nuclear weapons threatened to kill everyone on the planet. Nuclear power plants were prone to meltdowns that released deadly radiation or sent plumes of it into the environment, poisoning the land or travelling thousands of miles to cause millions of cancers over generations and displacing millions of people from their homes.

We regarded nuclear technology as frightening, invisible, deadly and all-powerful. In our minds, it went against everything that could be considered good and natural. When men in power chose it as a weapon, it left cities in rubble and spread vast clouds of radioactive fallout around the globe. Exposure to subatomic particles in a nuclear accident could rip apart the delicate structure of living cells, leading to mutations, diseases and death. Besides the well-known accidents at Sellafield, Three Mile Island, Chornobyl and Fukushima, the promise of carbon-free nuclear power was shrouded for us by the unsolved problem of how to store the radioactive waste, the risk of breeding components for even more nuclear weapons, and the escalating cost of decommissioning older power stations.

These were the stories I had been told by my community, my friends, my heroes, the books I read and the people I idolised. I believed them with every fibre of my being, enough to risk spending time in jail. We called it 'being on the right side of history', a familiar argument that would be later used in XR as well. Little did I realise about how wrong I was.

CHAPTER TWO
FOLLOWING THE SCIENCE

'Science makes people reach selflessly for truth and objectivity, it teaches people to accept reality, with wonder and admiration, not to mention the deep awe and joy that the natural order of things brings to the true scientist.'
Lise Meitner

'Any sufficiently advanced technology is indistinguishable from magic.'
Arthur C. Clarke

An earlier story

I had an unusual childhood. I was not born into a financially wealthy family, although we always had a roof over our heads and food on the table. Like all children of immigrants, my siblings and I grew up trying to balance two different cultures. But the biggest struggle I had as a child was of intense eco-anxiety, though we had no term for it at the time. My parents had different concerns – paying the bills and fitting in as migrants. Dad would tell childhood stories of being chased by Teddy Boys at weekends, who would beat up Asian kids for sport, and he was utterly puzzled by my nightmares of submerged cities. Back then hardly anyone seemed worried about what we called 'global warming'. But I was interested in the science, and what I learned disturbed me. So I did the only thing I could; with no understanding of my fears at home, I joined groups that shared the same worries, and threw myself into trying to save the world.

Fast forward twenty years, and suddenly everyone was talking about 'climate change'. We had XR to thank for this, and I was a member of the XR Media and Messaging team when we brainstormed the terms we wanted to popularise and came up with 'climate emergency', 'climate crisis' and 'net zero', amongst other catchy phrases that took the world by storm.

On the one hand, Andrew Neil was right that some of XR's claims were unfounded and damaging. On the other hand, it is a stark reality that the climate is changing because of human activity; and we have failed to act quickly to this threat. And then comes the greater issue – the environmental movement – those same people who claim to most understand the danger and are trying to solve the problem, have used it to promote a false solution: the 'living on the land' fallacy.

Since the Scientific Revolution of the sixteenth and seventeenth centuries, humanity has developed a systematic way to assess risk, which doesn't depend on individual guesswork or trial and error. Instead of each person having to determine for themselves whether an activity is safe, we rely on the scientific method: careful observation, experimentation, statistical analysis and peer review. This revolution, marked by the work of thinkers like Galileo, Bacon and Newton, laid the foundations for modern science and gave us the tools to evaluate risk objectively. Whether testing the safety of a new drug or designing a bridge, this evidence-based approach allows society to make informed decisions, grounded in data rather than personal perception.

The scientific method enables us to rely on a consensus of work by experts in any given field, which means that you and I don't have to spend decades measuring melting ice caps or monitoring hedgehog numbers to know that climate change is real and many species are on the decline. The IPCC, a scientific body of hundreds of researchers from around the world, has a scientific consensus of 99%, meaning there is an exceptionally high level of agreement on the data in its reports.

In the IPCC 2018 *1.5C Warming Report*, a chapter by Working Group III looks at mitigation methods for bringing down global greenhouse gas emissions, and concludes that a combination of nuclear energy, renewable energy, and carbon capture and storage (CCS) is needed to combat climate change. The clear scientific consensus, therefore, is that we need nuclear energy to decarbonise.

The reason wind and solar power are part of the picture but cannot meet our energy needs alone, is because they still need back-up, or a baseload energy source, on days without enough sun or wind. This comes from hydropower where it is geographically available, or from fossil fuels or nuclear energy. We know that fossil fuels contribute to global warming and air pollution, yet most countries still heavily rely on them. No one has succeeded in overcoming the laws of physics when it comes to wind and solar power – and not for lack of trying.

When I lecture on clean energy and tackling energy poverty, I'm often asked a variation of the following question: 'How do we stop consumption and development in order to protect the planet?' My answer usually draws blank stares, because it goes against the stories we've been told for years. We have been told to live with less, and that reducing personal carbon footprints will save us. I once wrote an entire book on how to live with a lower carbon footprint. But countless people have chosen to live like this, and the impact has been minimal.

Germany again serves as a good example here, because the country has been exemplary in implementing energy-saving measures aimed at improving efficiency across sectors. These include strict building codes demanding high insulation standards, retrofitting older homes, and promoting the use of energy-efficient appliances and lighting. The German government has also supported the rollout of smart meters to help households better track and manage their electricity use, alongside public campaigns promoting energy-conscious behaviour.

However, despite these commendable efforts, their overall reduction in energy consumption has been modest. Economic growth, increasing electrification and the persistent energy needs of a modern society, have counterbalanced many of the gains. This shows that while efficiency measures are important, they alone are insufficient to significantly cut overall energy demand without broader systemic changes or breakthroughs in clean energy supply.

So how do I answer the question? We got it wrong; we believed the wrong stories; living with less and halting growth aren't the answers to tackling climate change. Although we do need to decarbonise electricity grids and shift away from fossil fuels towards clean energy, achieving this demands more growth, not less. Resisting development won't save the planet, and it has not been proven to cut emissions either.

Sometimes, my remarks are met with anger – and I understand why. I too once believed that halting economic growth was the key to tackling climate change. For years, well-meaning activists have told us that our lifestyles are the problem: we must change everything immediately, stop using oil, eliminate plastic and give up driving. While adopting more sustainable habits has its merits, many of these ideas are, at best, oversimplifications. In recent years, this has surfaced in an ideology that has come to be known as 'degrowth'.

It is a tough pill to swallow. For writing my book on how to live with a low-carbon footprint to help save the planet, I was called 'Britain's greenest parent' by *The Telegraph* newspaper and 'an eco-pragmatist, happily heavy on evidence' by *The Guardian*. The mainstream press loves to champion the environmentalist argument of degrowth: living with less. And while I maintain that reducing our personal carbon footprints remains a worthy goal if it makes us feel good, and we should all strive to waste less, I realise that back then I was telling the wrong story. Even if we all make these lifestyle changes, it will not solve the problems we face. When it comes to combatting climate change, the focus on individual actions has, at best, been a misleading distraction from meaningful change. As well, it has shifted the burden onto ordinary people while ignoring the elephant in the room: how we power our lifestyles.

The good news is that, if we can stomach accepting different, or new, stories, there are solutions that have been proven to work. The Tooth Fairy, or a vision of a world powered by the elements alone, does have a magical Disney-like beauty to it, while picturing industry and industrialisation typically does not. But that's just because we haven't been very imaginative at telling those stories so far. After all, wind turbines and solar panels are still forms of technology, which come with their own environmental footprint. If we want realistic solutions to our environmental problems, we have to factor that in and make choices based on the available data. At the moment, all signs point to nuclear energy.

It might seem hard to believe that walking our current path could eventually lead to progress, since industrial developments have also been drivers of climate change. Yet evidence shows environmental improvements are already happening in many areas, and that – contrary to what we have been encouraged to believe – economic growth has actually been the driving force behind lowering emissions and improving air quality. If we truly want to save the planet and make life better for everyone, we need to stop believing the old stories and start understanding what works, so that we can tell better ones.

To sharpen our critical thinking and bring our judgement in line with the evidence, let's examine the leading solutions most prominently evidenced today for tackling our energy and climate challenges.

Updating the story:
1 – Reducing resource consumption

It's true that for the past 200 years, economic activity has led to an increase in carbon emissions. More recently, however, since the 1980s and largely thanks to using nuclear energy, many countries have reduced their emissions while continuing to increase GDP. Investment in renewables has further driven this growth. In 2016, seventy countries experienced a growth in GDP while also experiencing a run of at least five years[8] over which emissions decreased.

Using historical data, researchers[9] calculated that nuclear energy has by now prevented over two million air pollution-related deaths and 64 gigatonnes (Gt.) of CO_2-equivalent greenhouse gas emissions.

> 'On the basis of global projection data that take into account the effects of the Fukushima accident, we find that nuclear power could additionally prevent an average of 420,000–7.04 million deaths and 80–240 $GtCO_2$-eq emissions due to fossil fuels by mid-century, depending on which fuel it replaces. By contrast, we assess that large-scale expansion of unconstrained natural gas use would not mitigate the climate problem and would cause far more deaths than expansion of nuclear power.'

Perhaps surprisingly, we can see that as countries become richer, they actually become more environmentally friendly.[10] Air quality improves, water use becomes more efficient and fewer natural resources are required. This is true across developed countries, where the population has increased but resource consumption has fallen, including timber, water, metal, minerals and energy.

Much of this is driven by innovation and high environmental standards. There is a saying that a parked car in the 1970s was more polluting than a modern car driving at 100 mph today. This is largely because old automobile models were inefficient and heavy, which meant they required more fuel. Thanks to better designs and environmental regulations, even standard fuel cars have now become more efficient and less polluting. As we switch over to electric vehicles, further incremental progress will be made in this direction, particularly in reducing air pollution.

Although cars still carry an environmental cost, not having them would also have significant impact on the planet. *The Horse Association of America* calculated that 54 million acres of US farmland[11] was spared by the automobile in the 1930s, as the land was not needed as meadows for grazing horses. The US population has nearly tripled since then, yet hundreds of millions of acres of forests were saved from being cut down to make room to feed horses and store waste horse manure.[12]

I still don't own a car and have never learned to drive, and I confess that my own leanings towards the nature fallacy lead me to romanticise a world where horses replace automobiles as something idyllic and preferable. They seem natural, but that doesn't mean they are better for the environment. The data clearly tells a different story.

Cars aren't the only example – consider your smartphone. Innovations in phone design have drastically cut down on material use. Not long ago, a person might own a GPS device, a calculator, a camera, a landline phone, an alarm clock and more. Now, all those functions fit into a single device. While phones still require resources, the overall demand for mining and materials has dropped significantly now that one device replaces many. This kind of innovation fuels environmental progress, as technological efficiency often leads to substantial reductions in resource consumption.

Additionally, not having to purchase multiple separate products has made these technologies more affordable, allowing far more people to access them. In this way, technology acts as a powerful equaliser. As access expands, people's quality of life improves, helping them bridge the gap between themselves and more developed regions. Thanks to the smartphone, for example even in some of the world's poorest areas, many people can now connect to the internet, giving them unprecedented access to a vast wealth of knowledge. While seeming insignificant to those of us accustomed to instantly finding any information with a tap or voice command, for those without access to formal education, healthcare or other basic services, internet access can be truly transformative – even life-saving.

What can we do to reduce consumption and resource use? Again, the answer is counterintuitive, because of the stories we have heard for so many years. Through progress, i.e. promoting smarter design through innovation, and enforcing strong environmental standards alongside development, we can reduce how much raw material we need to dig out of the ground.

Updating the story: 2 – Reducing air pollution

The second solution addresses the major health threat of indoor air pollution, which involves an interim shift to cleaner fossil fuels – a step that may seem counterintuitive for those of us who have been protesting fossil fuels for years, but is the only transition that has been shown to work.

In many parts of the world, particularly in low-income and rural communities, cooking still uses traditional biomass fuels such as firewood, charcoal, animal dung or crop residues. While these materials are often freely available, their use carries the hidden cost of indoor air pollution. Household air pollution was responsible for an estimated 3.2 million deaths in 2020, and over 237,000 of these were children under the age of five.[13] These fuels lead to myriad health issues including non-communicable diseases such as stroke, ischaemic heart disease, chronic obstructive pulmonary disease and lung cancer, especially among the women and children who are typically responsible for household chores.

One of the cleanest and most efficient alternatives available today is liquefied petroleum gas (LPG) stoves – a technology with the potential to revolutionise cooking practices and improve public health. However, despite their benefits, LPG stoves remain financially and logistically inaccessible to many people who would benefit from them most.

The stoves work by burning liquefied petroleum gas, a fossil fuel stored under moderate pressure in liquid form for easier storage and transport. Compared to biomass stoves, LPG stoves produce far fewer air pollutants, including harmful particulates and carbon monoxide. They also cook faster and more efficiently, reducing the time spent gathering fuel or tending to the fire. From an environmental and health perspective, the transition from biomass to LPG represents a major step forward.

Cast your mind back to my alternate timeline – the woman coughing while cooking over the outdoor wood-burning stove who represents reality for millions of women who live in poverty around the world today. Yet, this problem is solvable, as the technology to transition away from dirty fuel stoves already exists. Outdoor LPG stoves are both cleaner and safer alternatives, but the sad reality is that only 60% of the world has access to clean cooking fuel. The only progress made in this area is when people attain higher income levels, meaning they can afford to switch to cleaner non-solid fuels such as ethanol and liquefied petroleum gas (LPG), which improves air quality significantly. This brings us back to considering economic growth and the need for development.

In the past, as an environmental activist, I admit that I disregarded the role petroleum gas could play in saving lives and improving health in the neighbourhood where my parents grew up, simply because gas falls under 'fossil fuels'. But, like renewables, not all fossil fuels are created equally.

Aversion to supporting fossil fuels in any way is one barrier against investing in LPG in low-income neighbourhoods, because people and companies don't want to be *seen* doing it. The same eco-conscious advocates who push for renewable energy transitions often reject LPG as a fossil fuel. It's easier to let perfect be the enemy of good when you don't live the reality of millions of people who cook over heavily polluting stoves. Yet, for billions of people still exposed to dangerous cooking smoke, LPG represents a critical intermediate step in a just and effective energy transition. It offers a cleaner bridge away from traditional fuels and towards healthier, more sustainable systems.

A further barrier is the upfront cost. Although the fuel itself may be affordable on a per-use basis, initial investment in the stove, gas cylinder, regulator and other equipment is substantial for households living on a few pounds a day.

The third obstacle is ongoing affordability. Traditional fuels, while inefficient and polluting, can often be freely collected from the local environment. The shift to LPG introduces a recurring expense that competes with daily survival needs like food, school fees or healthcare. For families with little to no disposable income, spending money on fuel that was previously 'free' can seem outpriced and irrational, even if its long-term health benefits are clear. It becomes evident that there is no single solution to these challenges, but lifting low-income neighbourhoods out of poverty by increasing development and economic growth is core to the broader solution.

Finally, there are social and cultural barriers. Traditional cooking methods are often deeply embedded in cultural practices and daily routines (consider my mother making *chai* in her beloved saucepan). Switching to LPG may require not just new equipment, but new cooking habits, meal timings and trust in unfamiliar technologies. Without local education and outreach, even a fully subsidised stove may not be used.

While LPG stoves offer a viable solution to the global indoor air pollution crisis, they are still out of reach for the most disadvantaged. Bridging this gap will require not only better technology and lower costs, but also thoughtful policy, reliable infrastructure and deep community engagement. Without such support, the promise of clean cooking will remain a luxury, rather than a universal right.

Even better, and looking past the interim measure of LPG, access to electricity allows the choice of much safer indoor gas and electric stoves, which greatly improves health and reduces risk. These options free people from exhausting and time-consuming tasks of gathering wood or hauling heavy LPG canisters in the scorching heat. With electricity comes freedom. It is why I have lectured across the world that 'poverty is energy poverty' and 'energy is life'; messaging that has since caught on and been used by many leading figures and nuclear influencers who have since entered the arena.

It sounds complicated and difficult, but it can be done. After all, even Britons once cooked like this, before we had electricity and before the invention of gas cookers. In recent years, almost every country in the world has made progress in tackling indoor air pollution, and its global deaths have declined significantly since 1990. Despite continued population growth in recent decades, the total number of deaths from household air pollution continues to decline. That's thanks to increased wealth.

Outdoor air pollution is hazardous to our health and the environment, but it has likewise improved over our lifetimes. Due to air pollution, British men born during the 1980s, in the most coal-intensive parts of the country were, on average, almost an inch shorter as adults than those who grew up with the cleanest air. Now, thanks to a raft of national and European legislation to tackle pollution, researchers have found[14] that over forty years, total annual emissions of fine particulate matter, nitrogen oxides, sulphur dioxide and non-methane volatile organic compounds, all reduced substantially in the UK. According to the study:

> 'Mortality rates attributed to PM2.5 and NO_2 (nitrogen dioxide) pollutants that increase the risk of respiratory and cardiovascular diseases declined by 56% and 44%, respectively, in the UK over the 40-year period. The estimated mortality rate related to pollution from ground-level ozone (O_3) – which can damage the lungs – fell by 24% between 1990 and 2010, following a significant rise in the 20 years prior to that.'

While there remains work to be done to improve air quality in industrialised nations, the progress made so far should not be overlooked. Thanks to stricter environmental standards, those of us in high-income countries now enjoy some of the cleanest air in the world – *thanks to* and *alongside* economic growth.

The same story can be seen throughout the developed world. In the US, thanks to the 1963 Clean Air Act (which was further updated in 1970), atmospheric sulphur dioxide dropped to levels that had previously not been seen since the first years of the twentieth century. Between 1980 and 2015, total emissions of six principal air pollutants decreased by 65%. When lead was banned from paint and gasoline, between 1976 and 1999 the concentration of the element in the blood of young children dropped by more than 80%. Researchers[15] found that this resulted in higher IQ levels in American children, less violent crime, and fewer unwanted pregnancies. Also, six common air pollutants[16] in the US have declined by 77% since 1970, despite the country experiencing continued economic growth. During this time GDP increased by 285% and population by 60%. Economists summarise[17] that in the US, 'changes in environmental regulation, rather than changes in productivity and trade, account for most of the emissions reductions'.

Expanding clean energy infrastructure is essential, not only for improving air quality but also for cutting emissions and minimising resource use. When construction is completed, Britain's new nuclear power plant, Sizewell C, will produce 3.2 GW of electricity, releasing virtually no emissions once it comes online. Compare this with the 2.6 GW produced by the Drax biomass power station in Selby, North Yorkshire: by burning 27 million trees every year, Drax is the UK's biggest emitter of carbon dioxide, as burning wood creates 18% more CO_2 than burning coal.[18] Burning wood in Britain generates 15.6 Mt of CO_2 emissions each year, of which 13.3 Mt alone comes from Drax. It is, essentially, the world's largest wood-burner.

Most of the wood burned at Drax is imported from North America. As explained in *The Guardian* newspaper, Drax is essential for 'when weather conditions preclude significant wind and solar power generation', as a back-up source of energy generation. Thanks to its classification as 'renewable', Drax also receives government subsidies.[19]

To meet targets that favour 'renewables' over 'clean energy', some governments are investing heavily in *only* renewables rather than choosing the most effective options to reduce emissions. This confusion arises because the terms 'renewable' and 'clean energy' are often used interchangeably, a misconception largely influenced by the environmental movement and what it considers to be natural versus human-made. However, as demonstrated by biomass, 'renewable' does not automatically mean 'clean', and as demonstrated by hydropower, it does not automatically mean 'safe' either.

Consider it this way: the alternative to building Sizewell C nuclear power plant would be burning the equivalent of thirty-three million trees annually. That's more than one tree every second. Nuclear energy isn't just better for the environment in the short term, it also drastically reduces outdoor air pollution. Instead of depending on polluting fuels simply because they're classified as 'renewable', we need to focus on building truly clean energy, and for this we need to include the nuclear technology that genuinely helps us breathe easier.

The harsh reality is that ill-informed political decisions, often driven by activist pressure, have deadly consequences. As a result of Germany's nuclear power phase-out, the country began burning coal again,[20] and a study[21] found that the resultant air pollution is killing an extra 1,100 people a year. Japan, too, shut down their nuclear power plants following the Fukushima meltdown (although Japan recently reversed this decision) and a study[22] found that if both countries had reduced fossil fuel power output instead of nuclear energy, they could have prevented 28,000 air pollution-induced deaths and 2,400 Mt CO_2 emissions between 2011 and 2017.

Of course, I wasn't aware of these facts when I was protesting against nuclear energy. Moreover, much of this research is recent, as the full environmental impact of nuclear shutdowns only became clear after the decisions had already been made.

Science eventually countered the activist storytelling, but science is also notoriously unimaginative at telling new stories. Data alone doesn't narrate a story. Only people can do that.

The environmental movement made a further mistake in ignoring the impact of air pollution. In my many years of active involvement in the movement, I have seldom heard air pollution discussed, despite it undeniably being one of the most critical environmental issues affecting all life on Earth. Many deaths caused by air pollution are preventable. Had we focused our efforts earlier on tackling this problem instead of opposing clean energy solutions, such as nuclear energy, we could have saved countless lives and mitigated some planetary warming.

Updating the story:
3 – We are not too many

Another belief closely tied to environmentalism is that people are inherently wasteful, and thus, fewer people would mean a healthier planet. The overpopulation myth cannot be separated from the environmental movement, as traditional environmentalism was founded upon it. In 1798, the English economist, Thomas Malthus, predicted[23] that a (so-called) 'overpopulation' would inevitably lead to famine as there would be too many mouths to feed.

The fear of overpopulation has long shaped global development policy, particularly in the mid-to-late twentieth century. Now widely recognised as an oversimplified and often inaccurate narrative, the idea that population growth alone would cause famine, poverty and environmental collapse drove some of the most coercive population control policies in modern history. Nowhere was this more evident than in India during the 1970s, where concern over population size, fuelled by international pressure and flawed assumptions, led to widespread, often forced, sterilisation of millions of people.

The global overpopulation panic was largely driven by

Western thinkers and policymakers. In 1968, biologist, Paul Ehrlich, published *The Population Bomb*, a book that predicted mass starvation and global catastrophe unless drastic measures were taken to curb population growth, particularly in developing countries. India was frequently portrayed as the poster child for this crisis – a nation of fast-growing numbers, struggling to feed itself. These perceptions, though largely exaggerated and devoid of nuance, found fertile ground in global policy circles.

As a result, international development agencies such as the World Bank and the US Agency for International Development (USAID) began tying financial aid to population control programmes. India, which was reliant on foreign support during a period of economic and political turbulence, was strongly encouraged to take drastic action. What followed was a nationwide population control campaign that would reach its most extreme form during the Indian Emergency of 1975–1977.

India's government implemented an aggressive, and now notorious, sterilisation campaign. In 1976 alone, more than six million men were sterilised, many of them forcibly or without informed consent. Some were lured with small cash payments or promises of food and employment, others were rounded up, coerced, or operated on in unsafe conditions. The poorest and most marginalised people bore the brunt of this programme, further entrenching social inequality and public distrust in government services.

These policies were justified by arguments from the aforementioned Western academics but based on the incorrect premise that population growth was the primary barrier to progress and wellbeing, rather than poor governance, lack of infrastructure, or uneven access to education and healthcare. By focusing on population numbers instead of human welfare, the sterilisation campaign overlooked the complex realities on the ground, and in doing so, caused widespread harm.

This period's legacy remains a powerful reminder of how policy shaped by fear, rather than evidence or ethics, can lead to serious violations of human rights. India's experience demonstrates the danger of letting ideology and international pressure dictate public policy. Any conversation about population and sustainability should centre human dignity, informed choice and accurate data. As we continue to face global challenges like climate change and inequality, it is vital that we learn from the past and reject simplistic narratives that place blame on people in this way.

Without a doubt, Malthus and Ehrlich were wrong. There was no population 'bomb' and no widespread famine caused simply by increasing numbers of people. Instead, life improved for millions, driven by innovation and development that people themselves initiated. This has been especially true in agriculture, where labour and capital grew faster than the population, boosting land productivity. Thanks to advances like the mechanisation of the Green Revolution, crop yields soared, food supplies improved, and economic development expanded in many previously underdeveloped regions.

Although land use and deforestation initially intensified to feed a growing population, we have likely now passed the point of 'peak deforestation'. Thanks to improvements in crop yields, the per capita demand for agricultural land has steadily declined: since 1961, agricultural land area has increased by only 7%[24] while the global population has grown by 147%. Emerging technologies like lab-grown meat promise to further reduce the land needed for livestock production. With great innovation comes significant environmental benefits.

We have also been repeatedly told that *we* are the problem – that the solution is simply to have fewer children. But this idea remains fundamentally flawed. It revisits the idea of humans being problems for nature rather than part of nature, as per the nature fallacy.

Contrary to common belief, larger populations are not necessarily more harmful to the environment. As populations grow, more people tend to move from rural to urban areas, which reduces the pressure on natural landscapes. The density of large cities brings environmental benefits because urban centres are often cleaner and more energy-efficient than smaller towns or rural areas. Cities are easier to heat, cool and navigate; and homes in urban settings typically emit less carbon dioxide than those in suburban or rural locations.

In fact, researchers have found that reducing the human population would have a negligible impact on climate change.[25] A child born today is expected to have a smaller carbon footprint than one born a few decades ago, and future generations are likely to shrink that footprint even further. Due to population momentum, global numbers will remain high for several generations, and by the time any decline occurs, individual emissions may already be minimal. The authors estimate that even significant population reduction would lower global temperatures by a mere 0.05°C, compared to stabilising our current numbers.

A strong case can be made that the growing population has actually fuelled environmental progress through increased human capital. In 2022, the world's population reached eight billion. Throughout history, exceptional individuals have driven technological and cultural breakthroughs, and the last 200 years have seen exponential growth in innovation and development. In 1823, the global population was just over one billion; today, with eight billion people, new advances in science, art, medicine and technology emerge every day. A larger pool of human capital is therefore not inherently detrimental; in fact, it can bring immense benefits, provided that many more people are liberated from poverty, and given the opportunity to live fulfilling lives and contribute to global progress.

Arguably, dense cities are essential for human progress, serving as vibrant hubs of innovation and transformation. Research[26] shows that metropolitan areas with high population density tend to be far more productive and inventive than rural or suburban regions. While the idea of living on the land in a secluded cottage with a few farm animals might seem idyllic and 'natural', it is not good for nature. Studies repeatedly find that such lifestyles have a larger environmental footprint than life in a compact, dense city.

Human capital drives innovation, and more people mean more minds to improve lives for everyone. I cannot express this argument better than the late statistician, Hans Rosling, who highlighted[27] the dramatic reduction in extreme poverty and shared how owning a washing machine transformed life for his financially struggling family:

> 'My mother explained the magic with this machine the very very first day. She said, "Now Hans, we have loaded the laundry, the machine will make the work. And now we can go to the library. Because this is the magic. You load the laundry, and what do you get out of the machine? You get books out of the machines. Children's books." And mother got time to read for me. She loved this. I got the ABCs, this is why I started my career as Professor, when mother had time to read for me… We really loved this machine. And what we said, my mother and me, "Thank you industrialisation. Thank you steel mill. Thank you power station, and thank you chemical processing industry that gave us time to read books"!'

For Rosling's mother it was a washing machine, for mine it was a stainless steel saucepan. Rosling's story is also my story, a shared story across generations, shared by countless individuals who have escaped poverty and their children who have witnessed it. I was not born yet when my parents bought their first washing machine, so I can only imagine how world-changing it felt for my mother, who would have realised, watching the drum spin, that she would never have to handwash clothes in a bucket of river water ever again. But I have seen her eyes light up when discussing some new (to her) technology that has invariably changed her life. From Tupperware to the refrigerator, she treasures it all, and throws nothing out. These items, while seen as vacuous materialism and consumerism by my former peers, are for her the gift of life; symbols of the energy poverty she once endured, left behind, and from which she freed her children.

I do understand why many environmentalists argue that humans are the problem, because we have caused climate change and various forms of environmental damage. However, over time, I have come to balance these views and understand that this perspective is simplistic and does not reflect the full reality of life. Being alive makes a dent on the planet – there is no getting around that. Every item that surrounds you has an environmental cost – if it wasn't grown, it was mined.

But this is where I have changed my thinking, because there is so much evidence that humans are capable of solving these problems as well. We have done so many times before, whether by reducing deforestation, saving species from extinction, or collaborating globally to protect the ozone layer. Does that sound like a species that wants to cause the planet harm? Reducing the population won't solve these challenges – we need people to do it. With people, comes co-operation, sensible environmental policies, and a commitment to innovation and discovery.

Now we find ourselves at a crossroads. Fossil fuels have powered our lifestyles for decades, but now that we know they harm our health and planet, we have a choice to make: do we give up dependence on these life-changing technologies, or do we continue our daily lives while choosing safer, cleaner alternatives?

By choosing the former, the following question naturally arises: can the world afford to allow countries like China and India to develop?

Sometimes I have to bite my tongue. Who are we trying to save the planet for, if not for all the life that inhabits it – including human beings?

The uncomfortable truth is that for many low-income countries significant fossil fuel use is a necessary step towards development and poverty reduction – the same step that industrialised nations once took – before they can realistically transition to cleaner technologies. As someone who actively protested all fossil fuels for decades, I have come around to this fact. On this subject, as with nuclear energy, I have been wrong; for years I protested *all* fossil fuel use on any grounds.

In the organisations in which I was active, we justified this by arguing for moving directly to 100% renewables, which would mean avoiding the growth step needed to escape poverty. But this is the foundational layer on which the house of cards begins to crumble. My peers also argued that poorer nations should accept the environmental setbacks involved in transitioning to cleaner energy as a necessary cost for humanity's future, which was one of those arguments that – given my heritage – I could never quite make peace with. I don't think that my friends meant harm by these views. Although they were arguing for harmful measures, this was the consequence of *not* living in poverty, making it impossible to conceive. Growing up with energy security and abundance makes it hard to grasp the reality that billions of people face on a daily basis, or to understand why they continue to burn fossil fuels, or that *other countries still need them in order to develop*.

Now, I urge those voices to reconsider, not only on scientific grounds, but on the basis of compassion and justice. I strongly disagree with environmentalists who suggest that poorer nations should remain undeveloped, or who romanticise poverty as somehow better; more natural, or even idyllic. It is none of these things. It is a harsh reality that millions suffer and, denying development, denies people basic human dignity.

Besides, it wouldn't achieve what we want anyway. As I already covered, when people strive to escape poverty, their energy use and associated carbon footprint inevitably rise *initially*, but once a high quality of life is achieved, individuals and societies are free to prioritise environmental concerns, reduce consumption and lower emissions.

If environmentalists truly care about minimising the impact of development, their aim should not be degrowth, but to support efficient rapid growth, paired with strong environmental standards, enabling cleaner outcomes sooner rather than later. In the meantime, it makes sense to champion the expansion of clean energy at home; particularly the nuclear energy that is essential to meeting our climate goals, as for wealthy nations like ours, there is no excuse. It is neither poverty nor lack of resources, but only the chains of outdated ideology, that prevent us from embracing the clean energy solutions we so urgently need. There is no valid justification for the rest of the high-income nations not to have decarbonised, like France did in the 1980s. Ideology is the only excuse, supported by convincing storytelling.

Frankly, I have come to find it abhorrent that in Germany, through burning coal again because they had phased out nuclear power plants, people died with the resulting increased air pollution, which also contributed to worsening climate change. Wealthy nations have this privilege; those already living in extreme temperatures do not. Either way, it causes immense harm to people and the planet. As an environmentalist, I simply cannot stomach it any more.

Let's give humankind a break. For too long, we've been sold the myth of 'overpopulation'; a fearmongering narrative that has been thoroughly debunked. Population decline is now a pressing reality in many countries. Japan and South Korea are experiencing record-low birth rates, while China's population is shrinking for the first time in decades. Across Europe, fertility rates have fallen well below replacement level, and even the United States has seen a steady decline in births. As younger generations delay or opt out of parenthood due to economic uncertainty, housing costs and changing social norms, governments are facing serious concerns about labour shortages, economic stagnation, and how to support rapidly ageing populations. Despite policy efforts to boost fertility or attract immigrants, many countries are struggling to reverse the trend.

Population decline brings new and serious challenges. Ageing populations strain healthcare systems and pension funds, and in countries like Japan, where there simply are not enough young people to care for the elderly, they face problems with lack of care and dignity. This demographic shift poses significant social and economic problems that cannot be ignored.

Japan's birth rate has been declining since the 1970s, and the country has the second-oldest population in the world,[28] where more than one in ten people are over eighty years old. Despite various government incentives aimed at encouraging higher birth rates, there is currently no indication that the trend of population decline is slowing.

Many environmentalists view population decline as a positive development as they feel that fewer people will reduce stress on the planet and lower energy consumption, but again, this is not true. Japan's experience tells a different story. Supporting its large elderly population requires substantial energy and, to cope, the country is increasingly relying on technology, especially power-intensive robots and data centres that run AI. As a result, energy consumption is projected not to fall with fewer births, but to increase by as much as 50%.[29]

It may initially sound dystopian, but using technology to address loneliness and healthcare needs amid a shortage of young people is no different than correcting eyesight with glasses or surgery, or protecting feet with footwear. It is undeniably positive. In Japan, robotics and AI are already helping to fill the gap in caregiving, supporting hundreds of elderly residents in care homes with only a handful of human staff. And thank goodness, because continued innovation in this field is likely to provide solutions for other countries facing similar demographic challenges, as many are not far behind Japan on the path to underpopulation.

Updating the story:
4 – Reducing greenhouse gas emissions

Back to a story that's rarely told: countries that pursue economic growth and development to overcome poverty are able to decouple their emissions from economic growth. Once they become wealthier, their focus shifts towards improving air quality and addressing environmental issues, so they implement stricter environmental standards and adopt cleaner technologies, which leads to a natural reduction in emissions and climate impacts.

This pattern has been observed in several high-income

Emissions are adjusted for trade. This means that CO_2 emissions caused in the production of imported goods are added to its domestic emissions – and for goods that are exported the emissions are subtracted.
Average incomes are measured by GDP per capita (except for Ireland, for which it is measured by GNI per capita).

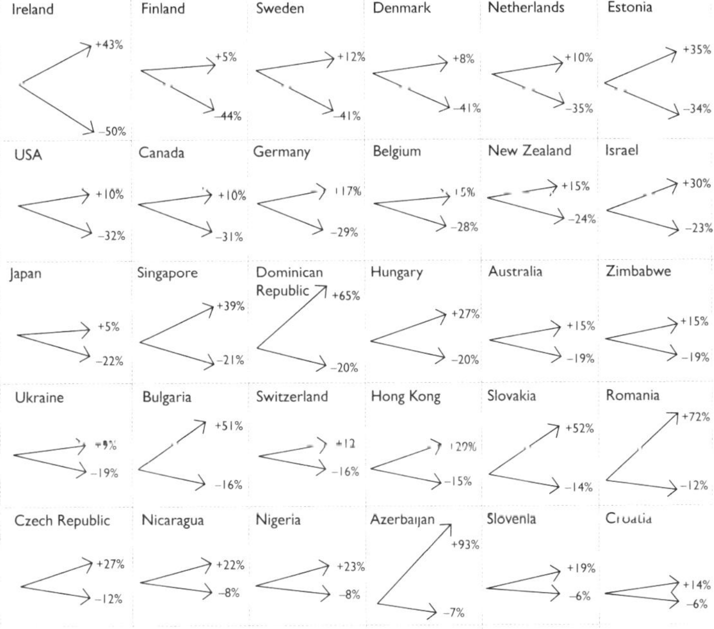

Data sources. Global Carbon Project & World Bank. There are more countries that achieved the same, but only those countries for which data is available and for which each change exceeded 5% are shown.

Decoupling: countries that achieved economic growth while reducing CO_2 emissions, 2005–20

countries, where air and water quality have improved even as GDP has continued to rise, and the United States provides one of the clearest examples. Since the implementation of the Clean Air Act in the 1970s, levels of pollutants have significantly decreased, while water quality in lakes, rivers and coastal areas has improved through enforcement of the Clean Water Act. These environmental gains have not hindered economic expansion even as emissions and pollution declined sharply.

The United Kingdom offers a similar success story. Over the

past few decades, the UK has drastically reduced its reliance on coal, improved air quality in urban areas, and invested in wastewater treatment to clean up its rivers. These changes occurred alongside consistent economic growth, supported by a shift towards services, finance and green technologies. Germany and Japan have also followed this path, using regulatory frameworks and innovation to address pollution while maintaining their positions as global economic powerhouses.

Even countries that experienced rapid industrialisation more recently, such as South Korea, have taken steps to reduce domestic air and water pollution. Although they still face environmental challenges, particularly due to their energy-intensive manufacturing sectors, these countries are showing that it is possible to clean up the environment as incomes and infrastructure improve.

Because this goes counter to the narrative we are used to hearing, it can feel untrue. The natural sceptical response is that emission reductions in wealthy countries are simply the result of offshoring their manufacturing to nations like China or India, effectively shifting emissions rather than reducing them. Thankfully, this isn't true. When looking at consumption-based emissions, which account for those embedded in imported and exported goods, there remains a decline.[30] While some emissions have indeed been outsourced, the impact is far less significant than often believed. Most emissions still originate domestically, within national borders. In fact, economic growth has been successfully decoupled from emissions in thirty-two countries,[31] each with populations exceeding one million.

Let me put it another way: economic growth and environmental progress are not enemies, but allies. As nations develop, they gain the resources and technology to address pollution and climate change effectively. Rejecting growth in the name of environmentalism overlooks this vital connection and risks hindering solutions that can truly improve both human lives and the planet. Embracing innovation, supporting sustainable development, and setting strong environmental standards, offer the most promising path forward; one where prosperity and a healthier environment go hand in hand.

CHAPTER THREE

THE POWER OF STORYTELLING

'The world cannot be understood without numbers. But the world cannot be understood with numbers alone.'
Hans Rosling

'You can't have civilisation without being able to communicate.'
Andy Weir, *Project Hail Mary*

It took me a while to come around to the numbers. They are, after all, just abstract figures to most of us. Thousands of people in a statistic don't carry the weight that one story does about a single person. We are storytelling creatures, and stories are how we understand the world, even when the numbers don't quite add up.

What if we've been focusing on the wrong stories?

I often wonder if climate change would be as severe today if more countries had embraced nuclear energy like France did in the 1980s. Between 1975 and 1990, in response to an oil crisis, France built fifty-two new nuclear reactors, successfully decarbonising its electricity grid while simultaneously developing world-class engineering and construction expertise.

Yet, despite the achievement, this story has barely been told. When I was first interviewed about my pivot from climate activism with XR to climate activism as a nuclear advocate in the early 2020s, I often offered France as a case study, but neither journalists, editors, podcasters or even news presenters knew about the Messer Plan. Similarly, talking about energy abundance over energy scarcity was taken as a controversial statement, and was

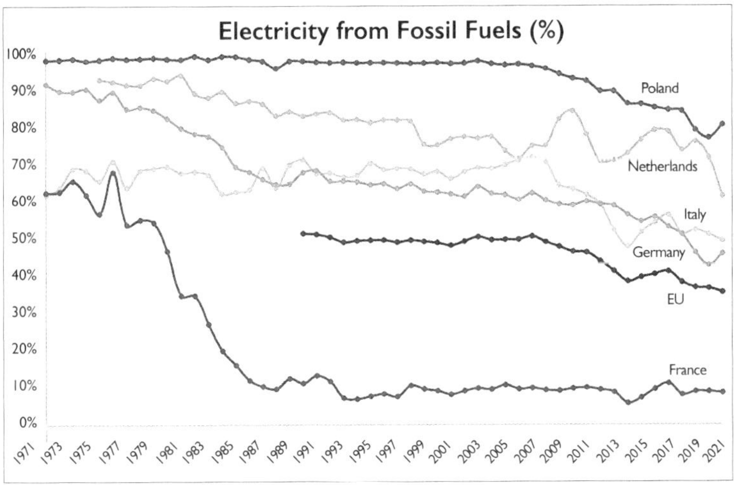

They have ideas: France reduced dependence on fossil fuels rapidly through building nuclear power plants

oftentimes edited out of my interviews and articles. The idea of an environmentalist saying these things was apparently mindblowing.

However, things also began to change. Telling this new story, which spread rapidly at the time, did excite people who had otherwise been worried about how we were possibly going to reach net zero with wind and solar power alone. The same story gained me attention in France, where I was invited to speak on a publicly broadcast, live energy debate opened by the current prime minister. It even reached Australia, where I was depicted on the news as the lone environmentalist standing up for nuclear energy and bucking the prevailing trend.

Yet even in France, there was intense debate. The French population was not immune to the old narrative that anything nuclear is bad and that the future should rely solely on renewables. The French government was experiencing pressure from not only the environmental movement, which campaigns heavily there to close existing nuclear power plants, but also the European Union.

Bowing to activist pressure, the French government made the political decision to close Fessenheim nuclear power plant in 2020. The closure fulfilled a pledge by President François Hollande to reduce France's reliance on nuclear from 75% to 50%, and it symbolised a shift towards renewables under pressure from anti-nuclear groups and neighbouring Germany. Regulators had cleared the plant to keep running, but it was shut down to signal progress on the 'energy transition', even though the loss of clean power meant greater reliance on fossil fuels. Since then, as the conversation has shifted, France has since pivoted back to nuclear, with fresh commitments to building new reactors.

The EU scenario is a telling example of where the story told by the numbers cannot represent the whole truth. Due to missing binding renewable energy targets, the EU attempted to fine France 500 million euros, despite the country consistently having some of the lowest emissions in Europe. Why? Because nuclear energy is not classed as renewable, even though it is clean and reliable. Ironically, Germany has avoided similar penalties, even though phasing out its nuclear power plants increased coal burning.

Either way, France's energy transition highlights a crucial lesson: ambitious investment in reliable low-carbon energy sources, like nuclear energy, can deliver significant environmental benefits, even if it doesn't perfectly align with prevailing policy trends. As climate change accelerates, it is imperative to learn from these successes rather than dismiss them for less effective solutions. Balancing innovation, pragmatism and science-based policy will be essential if we are to meaningfully reduce emissions and protect our planet for future generations. Instead of penalising France, we should ask why Germany has remained free of penalties, despite

Energiewende being the most expensive self-inflicted wound in modern climate policy.

Oral storytelling has been a cornerstone of human communication for tens of thousands of years. Long before the written word, knowledge, culture and values were passed down through spoken tales. Within this tradition, the elements of fear and drama are often amplified, not simply for entertainment, but because they serve essential cognitive and social functions. Modern science helps explain why these emotional ingredients are so effective.

From a neuroscientific perspective, stories that trigger strong emotions, such as fear, suspense or excitement, engage the amygdala, the brain region responsible for emotional processing. When the amygdala is activated it enhances memory encoding, by interacting with the hippocampus, which consolidates information into long-term memory. In other words, the more emotionally intense a story, the more likely it is to be remembered. This is particularly useful in oral traditions, where retention and repetition are vital for preserving stories across generations.

Fear plays a unique role in attention. Studies in cognitive psychology show that humans are evolutionarily wired to pay attention to potential threats, which researchers call the negativity bias. When a story includes danger, loss or suspense, it taps into our survival instincts, causing a rise in adrenaline and cortisol levels. These stress hormones sharpen our focus and increase heart rate, making us more alert and engaged. In a storytelling context, this ensures that the audience stays emotionally invested.

Similarly, drama amplifies conflict and high stakes within a story. According to narrative transportation theory, a highly dramatic plot allows listeners to become 'transported' into the story world. This immersive experience activates multiple brain regions, including those responsible for language, motor imagery and empathy. Listeners don't just hear the story; they *feel* it. This emotional simulation helps build empathy, cultural understanding and social cohesion; all essential outcomes of oral storytelling in traditional communities.

Together, fear and drama enhance the didactic power of stories. In oral traditions, tales often contain moral lessons or warnings. Cautionary stories about dangerous behaviour, whether they involve monsters, spirits or natural disasters, use fear to make abstract principles tangible and memorable. By dramatising the consequences, storytellers ensure that the audience internalises key lessons about survival, ethics or social norms.

The amplification of fear and drama in oral storytelling isn't accidental, but a biologically and culturally effective strategy. These elements engage the brain's emotional centres, sharpen attention, improve memory retention and deepen listener empathy. Through the power of heightened emotion, oral traditions have preserved human knowledge, values and wisdom across generations.

But what happens when stories lead us to fear new, life-saving discoveries or technologies? Nuclear energy is not the only demonised and misrepresented technology. Similar patterns of feelings affect many life-saving technologies, including childhood vaccinations, gene-editing (or as activists like to call gene-editing, 'GMOs') and modern day agricultural practices. Despite their proven benefits, these innovations often face resistance fuelled by fear and misunderstanding.

Misinformation about nuclear energy mirrors the myths surrounding vaccines: it is wrongly portrayed as unsafe, unclean, catastrophic and dangerously toxic, with high risk of harm to human health. This misconception is largely the result of a clever and harmful narrative crafted by the environmental movement, who have equated nuclear energy almost exclusively with 'toxic waste' and 'deadly radiation'. Meanwhile, the far more insidious waste from fossil fuels accumulates invisibly in the Earth's atmosphere. Yet, the term 'waste' is rarely invoked unless the topic is nuclear energy.

Nuclear energy saves lives. Many environmentalists, myself included, once believed – and some still believe – that nuclear power is unsafe, unclean or unnecessary. As a result, numerous countries have halted construction and even shut down operational nuclear plants, a decision that has contributed to millions of preventable deaths from air pollution and has worsened global heating.

In reality, opposing nuclear energy is like trying to put out a forest fire with a garden hose, while the fire truck sits idle.

It is safe to say that we got things wrong in the fight against climate change and air pollution. There was no grand conspiracy behind the widespread fear of nuclear energy, although many bad actors along the way likely provided fuel for the fire; rather, it stemmed from low scientific literacy combined with powerful storytelling across media and popular culture. This is, in essence, the definition of misinformation – stories that are spread whether or not with the intent to deceive. Stories that are repeated because we believe them to be true.

However, we cannot place all the blame on environmentalists. Where the environmental movement may have faltered in addressing climate change, the scientific community and the nuclear industry also share responsibility; their muted response allowed activists to dominate the story around nuclear energy for decades, while those in the know failed to attempt to tell better – factual – stories.

Misinformation has consequences, and has to be tackled head-on as soon as it rears its ugly head. In many cases, the nuclear industry failed to confidently counter people's increasing fears about nuclear power plants; fears that became further compounded by suspicions of corporate deception and a distrust of scientists. This is important because surveys indicate that people's concerns about nuclear energy are often tied to their perception of the industry, particularly regarding a perceived lack of transparency and trust.[32]

When I first challenged misinformation about nuclear energy, the industry itself reached out to discourage me, fearing that my efforts might backfire and worsen public opinion. While I understand their concerns, given the sometimes aggressive tactics of anti-nuclear activists, the stakes were too high for silence. Not only did the industry fail to engage the public when it mattered most, but it worked actively to dissuade me from speaking out. I wonder how many others were put off speaking about nuclear energy following its interventions.

Similarly, the climate scientist at the Green Party event knew he was being kept from answering the question about nuclear energy, and he could have insisted on speaking about it, but he chose not to, which meant that the 200 people in the audience only got to hear the lie that we don't need nuclear energy at all. These same scientists often lament the lack of progress on tackling climate change, while themselves failing to challenge false rhetoric when it is most needed, even when the opportunity arises.

I don't mean to berate anyone – this work is not for the weak. While the tide has shifted in favour of nuclear energy in multiple countries now, when I first pitched a book about it in 2021, I was turned down by major publishers who considered the topic to be 'too controversial' and something that the science was not settled on. Only now that so many people speak up about the need for nuclear technology have multiple new books appeared on the topic.

Clearly, I was in this same boat, sitting in the BBC studio and trying to be diplomatic despite my strong views against protesting nuclear energy. One thing to be said of being watched and laughed at by millions, is that you have nothing left to lose. I pivoted to championing nuclear energy after the disastrous television appearance, when I realised it couldn't possibly make my reputation any worse, and because I was fed up with having to keep quiet about it. What I didn't expect was just how many people would applaud these new stories, and even be inspired themselves to speak up for nuclear energy.

While I started as a lone voice in nuclear advocacy in the UK, I soon connected with supporters across the world who were relieved and excited that someone was finally telling better – and fairer – stories about nuclear technology. I realised I had found a new tribe, consisting of people from all walks of life – rich, poor, former environmentalists, scientists, artists and even celebrities. Through donations, I was able to tour the world giving lectures about nuclear energy and run public engagement events. I realised I had started something momentous and I needed to seize the moment.

For years I had lamented the lack of crossover between activism and the scientific community. When I had attended the March for Science in Bristol in 2017 I found it to be lacking, with few placards or chants and no music or 'rinkydink' – a protest staple composed of a bicycle with an attached cart and a large stereo that would lighten the mood around everyone it passed. So, in 2020, I co-founded a group called Nuclear for Net Zero. Following the lockdowns of 2020, which prevented us from doing public actions, in 2021 we hit the ground running, organising events around the UK and challenging mainstream myths of nuclear energy.

Later the group evolved into *Emergency Reactor*, with the help of director Robert Stone, Swiss entrepreneur Daniel Aegerter, and a former colleague from XR, Joel Scott-Halkes. Bringing Joel over to nuclear activism was another quiet success: in XR he had been the mastermind behind the Actions division of the group, organising large-scale protests in London.

Emergency Reactor held public engagement stalls in UK cities where we handed out hundreds of bananas and leaflets to passersby, answering their questions about nuclear technology. Everyone seemed to have a question for us, and 99% of people went away saying they felt positive about nuclear energy. I recall one striking moment in London where a long queue of people wanting to ask questions formed. We were asked about radiation, disasters, waste, *energiewende* and so much more. People were tired of being misled and just wanted to know the truth. We took it in turns to dress as Melty the Polar Bear to bring some humour to the events. Pro-activism like this had never been done before. We soon ran out of bananas.

We dropped banners across London, pestered politicians to speak up for nuclear, started petitions and open letters, and even dressed up as chimney sweeps and went to the Greenpeace headquarters in London to protest *their* protesting of nuclear energy, dumping charcoal on their doorstep to represent coal. At night we lit up London with messages projected onto government buildings demanding that they build more nuclear power plants – an action that became a trend, and has since been repeated by other advocates in Paris, New York and other cities.

I spoke at conferences and at rallies, wrote for magazines and reports. The more I wrote and spoke about nuclear energy, the more voices rallied to the cause. This felt like the activism I was used to – only the message to 'build more nuclear' was perhaps unexpected. There was too much for one person to do, and others began to take media interviews. I noticed that support for nuclear was starting to outweigh the hate mail and accusations. When I marched with nuclear workers on the climate march at COP26 in Glasgow, my friends were initially verbally attacked by environmentalists who felt we shouldn't be there. When I turned around to argue with them, they walked away. This is why my work had such impact – because of my story. While others who spoke up for clean technologies were bullied, pushed out, and ultimately left the environmental movement, I chose to stay. I had earned my stripes in this movement, and the most vocal anti-nuclear faction knew it.

At COP26 we staged a theatre piece where I dressed up as nuclear and married renewables, while coal tried to break up the wedding. The footage was used in a Germany documentary made for the public-service broadcaster, ARD, to examine their decision to quit nuclear energy.

Things began to change quite quickly. While that year (2021) we were not allowed into the climate section of the COP26 conference, just one year later, thanks to relentless pressure from advocates and growing public support, nuclear had won its place. The nuclear industry became bolder with its communication. I had coined multiple phrases that were being repeated all over the world, as well as talking points about nuclear energy that were, only a year ago, unknown to most people. I persuaded a former environmental campaigner to come back to public activism and attend rallies with us, co-authored an article with the musician José González, appeared in several global documentaries, and took every opportunity I was given to tell my story – in the press, in podcast interviews, in person. Although anti-nuclear activists continued to attack me all over the internet, accolades began to drown them out.

But the most powerful action I took was in persuading

two mainstream British newspapers to allow me to write their first positive articles about nuclear energy. Both articles were published in the same week in 2020[33], and they created a preference cascade where suddenly, countless commentators and editors decided to publish their own articles showing support for the technology. My arguments gained traction in Australia, where I was invited to discuss them on the news and later to go on tour to speak about nuclear energy. Suddenly, not only did people have permission to talk about nuclear in a positive way, but it also became a cool thing to do, partially because it was such a rebellious position to take at the time.

Finally, a new story had emerged – one that forged a new kind of environmentalism, and reshaped the world.

CHAPTER FOUR

THE DYSTOPIAN NARRATIVE

'Fear is the most valuable commodity in the universe… Fear is the most basic emotion we have. Fear is primal. Fear sells.'

Max Brooks, *World War Z: An Oral History of the Zombie War*

I don't particularly enjoy talking about disasters, but it wouldn't be fair to you to write a book about nuclear energy that doesn't at least touch on the three main disasters that made us so fearful. I understand, because they once filled me with a sense of dread as well, and I would have been grateful to learn that my worries were misplaced, had I known where to find accurate information about them.

If you're not interested in meltdowns, feel free to skip this chapter, but you may find that you too held misconceptions.

Chornobyl. Fukushima. Three Mile Island. These words used to fill me with terror.

In 2023, a train carrying hazardous material in Ohio, in the United States, derailed and caught on fire, leading crews to perform a controlled release of toxic chemicals. 2,000 residents were ordered to evacuate the immediate area. Officials were compelled to release vinyl chloride – which is associated with heightened risks of cancer[34] – into the air from five derailed rail cars at risk of exploding. Accidents like this are common in the United States; around 1,700[35] occur annually, and 25 million Americans live in zones vulnerable to derailments. Nuclear meltdowns are rare by comparison, and as we've seen, cause comparatively far less damage, but the fear of nuclear fallout runs deep, and even images like the above bring to mind the mushroom cloud of a nuclear explosion. We know that nuclear weapons have caused countless deaths along with decades of subsequent health issues, and we falsely associate nuclear *bomb* explosions with nuclear power plant meltdowns.

There have been three major nuclear power plant meltdowns since the advent of nuclear power in the 1950s – Three Mile Island in the United States (1979), Chornobyl in the former Soviet Union (1986), and Fukushima in Japan (2011). Of these, Chornobyl was the most severe, resulting in fatalities and long-term environmental contamination due to a reactor explosion. Fukushima involved meltdowns in three reactors following a natural disaster but caused no direct deaths from radiation. Three Mile Island was a partial meltdown with no casualties but had a lasting impact on public trust in nuclear energy. While other minor incidents have occurred in research or military reactors, these three are the only large-scale meltdowns in commercial nuclear history.

An interesting outcome of these disasters is how long-lived and terrifying the stories surrounding them have become. This is likely because how they were managed is as much part of the problem as the events themselves. Fear of nuclear power plants increased following accidents at Three Mile Island, Chornobyl and Fukushima – where the narrative was led by excitable journalists – subsequently shaped by anti-nuclear activists, and the public felt that their concerns were not addressed or heard by authority figures.

Most of us don't know what a meltdown is. I didn't, and I was frightened of radiation and potential meltdowns for many years. This fear was so paralysing that I protested against nuclear energy as though nuclear power plants were atomic bombs.

Of course, we should be prepared for nuclear meltdowns, but not so paralysed by fear that we make poor decisions when they occur. Every meltdown has a cause, whether technical, human or both, and understanding these causes is essential to making informed, rational decisions about the future of nuclear energy. So, let's examine the three most significant nuclear power plant incidents in more detail, not to fuel fear, but to learn from them.

The Chornobyl disaster is the most serious incident in the history of nuclear energy. The Chornobyl nuclear power plant comprised four Soviet-designed RBMK-1000 nuclear reactors, a design that is universally recognised as flawed[36] because it had a positive void coefficient, meaning that as the reactor's coolant boiled and temperature rose, the reactor's power output could increase uncontrollably, potentially leading to a dangerous power surge. The accident occurred in 1986 during a safety test. Inexperienced operators made a mistake: they dropped the reactor's power output to near zero, then tried to increase it rapidly, causing a thermal runaway. This led to a massive power surge – at least ten times the reactor's maximum output. The sudden spike instantly turned the cooling water into steam, resulting in a steam explosion, similar in force to those that could sink a ship during the age of steam. Compounding the problem, the reactor's safety systems had been manually disabled to carry out the test.

The situation in Chornobyl was made worse by the authorities choosing not to immediately evacuate the area, but waiting until around thirty-six hours after the accident had occurred. By that time, residents were already complaining about vomiting, headaches and other signs of radiation sickness. Nor were they immediately or adequately treated for radiation poisoning.

The RBMK[37] is one of the oldest commercial reactor designs still in operation. Following the Chornobyl disaster, its most critical design flaws were addressed, and upgraded RBMK reactors have operated without major incidents for over three decades. However, no new RBMKs are being built today, as modern reactor designs are significantly safer, more efficient, and better suited to meet today's energy and safety standards.

Chornobyl was a tragedy. The initial explosion at the Chornobyl nuclear power plant resulted in the death of two workers. Twenty-eight firemen and emergency clean-up workers died in the first three months after the explosion from acute radiation sickness, and one died of cardiac arrest.[38]

Compounding the tragedy, some doctors in Europe incorrectly advised pregnant women to undergo abortions following the accident, because they misunderstood radiation exposure. Robert Gale, a haematologist who treated radiation victims after the accident, estimated[39] that more than one million abortions were undertaken in the Soviet Union and Europe as a result of bad advice from doctors. Misinformation can be deadly. As I've highlighted, fear of nuclear energy is often more harmful than nuclear energy itself.

Informed and competent management is important in the aftermath of a meltdown. A report found that people in the area suffered a paralysing fatalism due to myths and misperceptions about the threat of radiation[40]. Heightening this perspective, '… individuals in the affected populations were officially categorised as "sufferers", and came to be known colloquially as "Chornobyl victims".'

Due to sensationalised media coverage and the rapid spread of misinformation, many people are unaware that the officially confirmed death toll from the Chornobyl disaster was less than 100 people.[41] They expect the number to be in the tens of thousands. Although there is some debate regarding potential long-term health effects, those estimates still suggest numbers in the thousands.

Again, I don't want to downplay this tragedy, but we do know that a disaster of this magnitude involving a nuclear reactor will not happen again, since the reactor design is no longer in use, and due to much stricter regulations which have been implemented since then. While tragic, nothing is zero risk, and no energy source is entirely without risk. According to data, the death rates per terawatt-hour (TWh) of electricity produced are approximately 0.04 for wind power, 0.02 for solar, and 0.03 for nuclear. Even combined, these figures remain remarkably low, especially when compared to fossil fuels. Clean energy does cause harm, but far less than many alternatives.

But I know it doesn't feel that way, and it took me a long time to come around to accepting it myself. I spent many years of my life surrounded by people who convinced me that tens or thousands of people had died because of Chornobyl (this is false) and that similar meltdowns would occur multiple times in my lifetime (they have not). A classic case of being more influenced by a frightening story than by the quieter, less dramatic story told by the numbers.

Next we have Fukushima. The nuclear disaster occurred in 2011 at the Fukushima Daiichi Nuclear Power Plant in Japan, and was caused by the Tōhoku earthquake and tsunami, which was at the time the most powerful earthquake ever recorded in Japan. The earthquake triggered a powerful tsunami, with thirteen-metre-high waves overwhelming the nuclear plant's sea wall and shutting down its emergency back-up from diesel generators. Back-up batteries kept the plant's cooling pumps running for some hours but, amidst the destruction of the tsunami, the plant's crew were unable to restore power before the batteries ran down so that, one by one, the reactors boiled dry and overheated, releasing hydrogen from the reactors into their enclosing buildings; they ignited in explosions, throwing up small but ominously mushroom-shaped clouds that were seen on live TV around the world. Inside the reactors, the cores melted down and a modest amount of radiation was released; enough to require rigorous safety precautions but not apocalyptic. Had the accident occurred in any of the other industrial plants devastated by the natural disaster it would barely have been news but unfortunately, being a nuclear plant, it was elevated to the status of a major international disaster.

In the hasty evacuation that followed, 573 stress-related deaths were recorded.[42] The poorly-managed evacuation process caused people to panic, which led to accidents, injuries and deaths. No one died due to the actual meltdown, however.

The most significant response to events in Fukushima was from Germany, who, despite not being vulnerable to tsunamis, decided to close their entire nuclear power fleet. As a result, their coal consumption has skyrocketed, and a report[43] found that 630 people in Germany died from coal-related illnesses in 2013 alone. Another paper[44] concluded that this air pollution is now killing an extra 1,100 people a year – vastly more than were harmed by the Fukushima nuclear power plant meltdown. And that's before we factor in the toll from increased global heating, of which one of our most powerful mitigating tools is nuclear energy.

Japan too shut down their nuclear power plants in the wake of the disaster (although this decision has recently been reversed). A study[45] found that if both countries had reduced fossil fuel power output instead of nuclear energy, they could have prevented 28,000 air pollution-induced deaths and 2400 $MtCO_2$ emissions between 2011–2017.

The idea of saving 28,000 lives, versus that of nuclear meltdown, does not make an exciting story. But in this case, the numbers ought to matter, because they represent real people who deserve better.

This is not the only example of how fear of nuclear energy had led to more harm than the meltdown itself. Following the disaster, there were multiple investigations, and one initially found a small increase in thyroid cancer rates amongst the surrounding population. The United Nations Scientific Committee on Effects of Atomic Radiation[46] and the World Health Organisation[47] concluded that this increase was due to better screening for thyroid cancer, not due to an actual increase from radiation exposure. But it added weight to the existing story – I was told that many people died from cancer due to the meltdown.

The family of one plant worker, who died of lung cancer, sued the government for damages and won, but the cause of his death is disputed,[48] as lung cancer is not a likely outcome of a nuclear meltdown, and other workers were not affected. In this case, even the legal system believed the compelling myths over the numbers.

Most people don't realise that while many people died due to the actual earthquake and tsunami, and some from the poorly handled evacuation, no one died because of the meltdown of the Fukushima Daiichi power plant. The real negative consequences came from a panicked reaction to the disaster.

Finally, we have Three Mile Island. The Three Mile Island accident, in 1979, was the partial meltdown of a nuclear reactor in Pennsylvania. It is the most significant accident in US nuclear power plant history and was caused by a cooling malfunction coupled with poor instrumentation of the plant, causing part of the core to melt in the TMI-2 reactor and destroying it.

Some radioactive gas was released due to the accident, but not enough to cause any dose above background levels of radiation beyond the plant perimeter. More than a dozen major independent health studies of the accident found[49] that:

> 'The approximately 2 million people around TMI-2 during the accident are estimated to have received an average radiation dose of only about 1 millirem above

> the usual background dose. To put this into context, exposure from a chest X-ray is about 6 millirem and the area's natural radioactive background dose is about 100–125 millirem per year for the area. The accident's maximum dose to a person at the site boundary would have been less than 100 millirem above background.'

After thousands of environmental samples of air, water, milk, vegetation, soil and foodstuffs were analysed, researchers concluded that the release of radiation had 'negligible effects on the physical health of individuals or the environment'. Most people don't realise that there were no injuries, deaths or direct health effects[50] caused by the meltdown at Three Mile Island.

What happens after a nuclear meltdown varies by incident, but rebuilding life in the aftermath is always a challenge. Some sites, such as Chornobyl, remain largely abandoned due to contamination, though parts of the area are now accessible. Radioactive isotopes like Strontium-90 and Caesium-137 still linger in the environment, but their levels have decayed significantly and are considered tolerable for limited exposure. In fact, some former residents of the Chornobyl exclusion zone have chosen to return and live there, despite elevated background radiation. These levels are not immediately fatal, and the health risks from low-level, long-term radiation exposure are generally much lower[51] than from a single, high-dose event.

Nuclear technology does not exist in a bubble. Following the disasters, strict guidelines and monitoring have been put in place for nuclear sites. Nuclear safety is overseen internationally by the International Atomic Energy Agency (IAEA), which sets global safety standards, conducts peer reviews, and promotes co-operation. Industry groups like the World Association of Nuclear Operators (WANO) also work to improve safety through peer reviews and sharing best practices. Enforcement is handled by national regulators, such as the US Nuclear Regulatory Commission (NRC), France's Autorité de Sûreté Nucléaire (ASN), the UK's Office for Nuclear Regulation (ONR), Japan's Nuclear Regulation Authority (NRA), and China's National Nuclear Safety Administration (NNSA), which licence, inspect and monitor nuclear facilities within their borders. Together, these bodies create a multi-layered system of oversight intended to prevent accidents and ensure safe nuclear operations.

If only the same learning had taken place for fossil fuels.

I don't want to downplay the tragic loss of life and harm that has resulted from nuclear meltdowns. Every death matters, and we must continue to do everything possible to prevent future accidents and protect both human health and the environment. But we also need to be honest, that these learnings have already taken place, and that no activity in life is entirely without risk. Whether it's driving a car, heating our homes, or even walking down the street, risk is a part of daily existence. The goal should be to manage and minimise these risks while weighing them against the alternatives, which in this case is fossil fuels.

I always say that when the message is muddied, that isn't the fault of the audience, but of the spokesperson with the message. In the case of nuclear meltdowns, communication could have been better – faster, clearer and more transparent.

At a press conference following the Three Mile Island incident, a vice president from the plant's owning company appeared visibly frustrated by journalists pressing for details about what had happened and what steps were being taken. At one point, he snapped, 'I don't know why we have to tell you every single thing we do!' This handed a gift to anti-nuclear activists. Clear, transparent communication is crucial to maintaining public trust with any technology, but unfortunately, no one took the lead in providing that reassurance. As a result, nuclear energy became increasingly burdened by heavy regulations, leading to longer construction times and skyrocketing costs for new plants. Meanwhile, the coal industry had fewer safety and environmental regulations than nuclear energy.

It became a vicious cycle. Fear of negative press and anti-nuclear criticism fuelled perception that the industry lacked transparency. Ironically, this caused the industry to shy away from the very clear communication needed to prevent public backlash. The Fukushima Daiichi meltdown exemplified this, as the Japanese government and nuclear operators' failure to communicate openly and promptly, significantly shaped public reaction to the disaster.

According to research by the Sasakawa Peace Foundation:[52]

> 'Overseas, detailed information was being provided promptly to the general public, but because the relevant authorities failed to adequately explain the differences in how the forecasting systems were being applied within Japan and abroad, it led to a great deal of distrust and confusion among the public.'

The Fukushima Nuclear Accident and Crisis Management alliance additionally recorded:

> 'As a result of the Fukushima accident, the public was suddenly asked to make their own decisions on a situation – low-dose radiation exposure – on which even the scientists are unable to conclusively decide. Because it was vague as to whether the government's explanation was based on what was accurate scientifically or on what was correct for society, as a result many in the general public were distrustful of the government and tried to get "correct" information themselves.'

It's no surprise that the Fukushima meltdown sparked such a fierce backlash against nuclear energy in Japan. However, Japan's decision years later to reverse course and embrace nuclear power again, highlights the stark contrast between the initial fear-driven response, rooted in a lack of clear information and leadership, and the current recognition of nuclear energy's vital role in maintaining a stable power supply.

People are swayed by emotions more than by facts. Ensuring effective communication is crucial to sustaining support for any endeavour. As my efforts to change perceptions of nuclear energy gained momentum in the early 2020s, the industry and other advocates were also evolving their approach, and support for nuclear grew rapidly.

We could ask, what if the nuclear sector approached disaster management differently? A useful case study would be the Challenger disaster. On 28 January 1986, Space Shuttle Challenger broke apart seventy-three seconds after lift-off, killing all seven crew members. The launch was the most-watched live television event in US schools, NASA having arranged a satellite broadcast onto TV sets into schools specifically for the event. Horrified teachers were unable to find words to console an estimated 2.5 million pupils[53] who watched as the tragic footage was cut from their screens. The disaster was also witnessed by thousands of people attending in person at Cape Canaveral, Florida.

The disaster was devastating, claiming more lives than Three Mile Island and Fukushima combined – both of which were incidents with no fatalities. So why didn't this tragedy bring about decline of the space industry?

In an instance of exceptional, forthright communication,

President Ronald Reagan immediately addressed the nation following the disaster to make the now-famous speech:[54]

> 'And I want to say something to the schoolchildren of America who were watching the live coverage of the shuttle's take-off. I know it is hard to understand, but sometimes painful things like this happen. It's all part of the process of exploration and discovery. It's all part of taking a chance and expanding man's horizons. The future doesn't belong to the fainthearted; it belongs to the brave. The Challenger crew was pulling us into the future, and we'll continue to follow them.'

He added context, explaining that occasional disasters are part and parcel of the process of exploration:

> 'There's a coincidence today. On this day 390 years ago, the great explorer Sir Francis Drake died aboard ship off the coast of Panama. In his lifetime the great frontiers were the oceans, and an historian later said, "He lived by the sea, died on it and was buried in it." Well, today we can say of the Challenger crew: Their dedication was, like Drake's, complete.'

Reagan took charge and rose to meet the demands of the moment: addressing a tragedy, assuring people, emphasising that this was a rare event rather than the norm, and adding that it would not deter the country's commitment to its goal of exploring space. He also heavily emphasised the importance of transparency – a great way of maintaining public trust:

> 'I've always had great faith in and respect for our space

> program, and what happened today does nothing to diminish it. We don't hide our space program. We don't keep secrets and cover things up. We do it all up front and in public. That's the way freedom is, and we wouldn't change it for a minute. We'll continue our quest in space. There will be more shuttle flights and more shuttle crews and, yes, more volunteers, more civilians, more teachers in space. Nothing ends here; our hopes and our journeys continue.'

In short, he demonstrated exceptional leadership when it was needed most. Notably, this level of communication has never been seen in the wake a nuclear disaster.

Meanwhile, the space industry continues to thrive despite its significant environmental impact, high financial costs and ongoing risks. Like nuclear meltdowns, the Challenger disaster was officially attributed to human error. Yet, unlike with nuclear incidents, there were no mass protests against the space sector, nor did governments shut down their space programmes, even after a second tragedy with the Columbia shuttle in 2003. When people talk about space, they don't immediately recall these disasters in the way Fukushima, Three Mile Island and Chornobyl are usually the first thing mentioned in conversations about nuclear energy. Unlike in the nuclear debate, Reagan never ceded the narrative to fearful activists. There's a lesson in there regarding public engagement, one that I hope continues to catch on.

The ability to weave a compelling narrative has long been essential to our evolution.[55] Research finds that, throughout human history, skilled storytellers have been the most preferred social partners.[56] Even when we roamed the land without static homes and all the other perks of civilisation, storytellers had the most offspring. Pause for a moment and let that sink in. It wasn't necessarily the people who provided food, or shelter, or protection from predators who had the most babies. It was the storytellers.

It's easy to understand why. Storytellers have long been the heart of human connection, weaving tales that warm the soul on cold nights. They capture human imagination, play with emotions, creating space in a single sitting for grief, anger, fear, love and more. Through their stories, they passed down values to children, long before formal schools or temples existed. They gazed at the stars, naming constellations after gods, using the ever-shifting night sky as a living canvas to teach timeless morals and lessons.

Authors, writers, actors and directors – modern day storytellers – remain deeply revered today. This fundamental human need to be transported through stories hasn't changed. While we no longer gather around the fireplace, those who ignite our imaginations through words and images continue to captivate us.

Over time we have forgotten the old stories. We still use the Greek names for constellations, but no longer know the stories behind them. We retell Shakespeare but appreciate that much of it requires cultural context to be fully understood. We use the days of the week, but few of us know the origins of their names (the Romans named them after their gods, corresponding to the five known planets plus the sun and moon, which they also classed as planets).

You might argue that those old stories are irrelevant now, out of place in today's world. But the same could be said of the story we've been telling about nuclear energy for decades. The value of revisiting nuclear disasters isn't to scare people – it's to show what *didn't* happen, and more importantly, what we've learned. That learning has made nuclear energy so safe that it now sets a standard for other industries to follow.

If we can release ourselves from the grip of ideological anti-nuclear rhetoric, we're free to ask new questions: What are the stories of today? Where are the narratives that bring us together and help us understand our place in the universe? Which stories drive out fear and illuminate the path forward?

And can humankind thrive without them?

CHAPTER FIVE

FACTS DEPEND ON FEELINGS

'Two roads diverged in a wood, and I –
I took the one less traveled by,
And that has made all the difference.'
Robert Frost

It seems like a lifetime ago now, but when I first started out in nuclear advocacy, at least thirty existing nuclear advocates contacted me within a few months to criticise my communication strategy. They all came from technical backgrounds and felt that I needed to explain more of the science. (Which they had been doing, but clearly hadn't been gaining much traction within public discourse.) Some even went so far as to tell me not to use slogans at all, proclaiming, 'We are not activists!' I soon came to understand why the narrative around nuclear energy had not changed or been effectively challenged since the 1960s.

Society has a problem with how scientific communication is shared, which is that there appears to be little crossover of the scientists and the storytellers. The latter – artists, directors, writers – understand and speak to humankind in a way that we understand. We read their books, watch their films, and may even subconsciously believe fictional narratives to be true, because they tell the stories so well. The scientific group wants us to appreciate numbers the way they do, and frequently speak down to us when we don't understand them.

In Science Communication this is known as the Deficit Model: the idea that people reject or misunderstand science because they simply lack the necessary information, and that the solution is to 'fill the gap' by providing them with more facts. In other words, it assumes the public is like an empty vessel that needs to be topped up with expert knowledge, and once they have that knowledge, they will naturally change their views or behaviours. It's as if the scientist is the jug, and the non-scientist is the empty cup – an old approach, and an ancient story.

This model is deeply flawed, for many reasons. In reality, people's responses to science are shaped by values, identity, trust and emotion as much as by facts. Someone may understand the science of climate change, for example, yet still reject it because of political or cultural beliefs. The model is also one-way: experts talk and the public listens, with no dialogue or trust-building. This is why many scholars now argue that the Deficit Model oversimplifies human behaviour, and overlooks the social context in which science is received.

I experienced this myself as a young environmental activist. A housemate, who was also a scientist, directly challenged me on my anti-nuclear viewpoint. Had they approached me thoughtfully and with kindness, like my friend John did many years later, I may have changed my mind much sooner. Instead, they patronised me, telling me that 'facts don't care about my feelings' – a common saying among scientists – and deepening my distrust in their institutions which, in my community, were merely different iterations of The Man. I had respect for science, but I felt disrespected and patronised by his approach, which only led me to lean into my anti-nuclear community further.

There is irony worth nothing here in that many large nuclear organisations, and most well-known scientists in similar fields, want everyone to understand *nuclear* science but themselves fail to accept, or even reject, the *neuro*science of storytelling, which is arguably the earliest science of humankind.

This is what Science Communication studies. Time and time

again, research shows that facts alone don't change people's minds, and that relying solely on the Deficit Model of trying to persuade people with facts and data usually falls flat. Instead, compelling storytellers who connect emotionally, even without purely factual content, hold far more influence. That is why we admire those masters who shape narratives that resonate. And even when we don't admire them, we can't help but use parody, or somehow repeat their messaging. I only have to say one phrase to make my point here: 'Make America Great Again'.

The challenge is to story-tell while remaining true to the data. Few have managed this successfully on a large scale, and we know their names: Carl Sagan, David Attenborough, Jane Goodall, Brian Cox, Emily Calandrelli, Bill Nye, Hannah Fry, Neil deGrasse Tyson and Hank Green.

But none of these communicators can cover all the topics available, and while many of them do spend time myth-busting, the topic of nuclear energy is not their main concern.

The problem is that when we fail to rise to this challenge, people fill the void with their own interests. A prominent example is the fuel behind the anti-vaccination movement – English doctor Andrew Wakefield. In 1998, Wakefield published a study in the prestigious scientific journal *The Lancet* that falsely linked the MMR vaccine, used to protect against measles, mumps and rubella, to autism in children. The paper claimed that twelve children developed autism-like symptoms shortly after receiving the vaccine. Though the study was small and anecdotal, it caused disproportionate, widespread fear and mistrust of vaccines. This single publication triggered a significant decline in vaccination rates in several countries, most notably the UK and US, and helped spark a modern anti-vaccine movement that continues to influence public health today.

Investigations later revealed that Wakefield's study was

riddled with ethical violations and scientific misconduct. He had falsified data, misrepresented the children's medical histories, and failed to disclose critical conflicts of interest, having been paid by lawyers preparing a lawsuit against vaccine manufacturers. As a result, *The Lancet* fully retracted the paper in 2010, and Wakefield was struck off the UK medical register for serious professional misconduct.

However, the damage was already done. Despite the retraction and overwhelming scientific evidence disproving the vaccine–autism link, many people continue to believe Wakefield's claims. There are several reasons for this. His narrative offered a seemingly clear explanation for the frightening and complex condition of autism, for which no single cause had been identified. Parents desperate for answers were particularly vulnerable to such claims. Moreover, distrust in pharmaceutical companies and government health agencies made Wakefield's story even more appealing to sceptics. Social media further amplified his message, giving it a new life even after being scientifically discredited.

Wakefield's legacy is a cautionary tale of how misinformation, especially when dressed in the language of science, can undermine public trust, cause lasting harm and endanger lives. In the years since, outbreaks of measles and other vaccine-preventable diseases have re-emerged in those communities where vaccine hesitancy took root. It is a powerful reminder of the need for rigorous scientific standards, transparency and critical thinking in our age of information overload. Ultimately, Wakefield cannot solely be blamed for what happened here. We have to ask how this paper ever passed peer review with such a tiny sample size, and why it was published in such a trustworthy journal. Despite the paper being retracted, it had already given Wakefield a voice that no one from the journal stepped up to counter.

I had countless conversations with vaccine-hesitant parents

about this paper when I was in green parenting circles and co-edited a hippy parenting magazine. They said they had more trust in Wakefield *because* his paper was retracted – it actually fitted into their inner narrative of 'us versus The Man' and of important information being covered up and kept from them. In this instance, they felt that The Man – scientific institutions – had tried to silence Wakefield because he was *right.* They did not accept my reasoning about the flaws in his paper; these were merely emotionless facts. Of course, they took their narrative of the 'underdog truthteller' directly from Wakefield's reaction to the retraction. Instead of it tarnishing his reputation, without public engagement and explanation, the retraction actually helped propel him to greater renown, and he continues to speak on this topic at large gatherings today.

Activists may lack factual evidence, whether with nefarious intent or through misunderstanding data, but they wield something far more powerful: the art of the storyteller. Scientists ignore this at their own peril.

Trust in those we rely upon to publicly communicate complex situations, which in Science Communication we refer to as 'communicator trust', is essential for dispelling misinformation and fostering a healthy climate narrative. My own journey of changing my mind about nuclear energy was far from easy but reflects this reality.

Whether we like it or not, activist messaging shapes our view of the world, influences what we believe, and affects how we think about global issues. The problem arises when this messaging is compelling but not grounded, particularly in science. For example, in Extinction Rebellion, phrases like 'mass extinction', 'tipping points', and slogans such as 'system change, not climate change', created a powerful narrative. These simple evocative concepts gave people a way to engage with climate change without using complex scientific jargon. Facts alone don't always persuade, but distilling ideas into clear relatable messages can be far more effective than expecting everyone to read and understand dense scientific papers. It's a dynamic that explains why people have for so long believed negative messaging about clean and reliable nuclear energy.

People often scoff at the social sciences and humanities subjects, while placing emphasis on the need for science, technology, engineering and mathematics (STEM). I think we need both, and that one cannot thrive without the other. Society does not run on facts alone. If the recent surge in anti-science sentiment hasn't made that clear, nothing will. Emotions drive human behaviour, so much so that gut feelings often matter more than data in our opinion-making. If we fail to tell compelling stories that celebrate scientific progress, stories that provide context beyond dystopian apocalyptic narratives, we are overlooking a crucial tool in the fight against rising pseudoscience. Storytelling isn't just art; it's strategy. And increasingly, it is a tool for survival.

CHAPTER SIX

THE SCIENCE AND TRUE STORY OF NUCLEAR ENERGY

'Hot water was civilisation.'
Terry Pratchett

Nuclear energy is produced by harnessing the energy released from atomic nuclei during nuclear reactions, most commonly through a process called nuclear fission. In fission, the nucleus of a heavy atom such as uranium-235 or plutonium-239 is split into two smaller nuclei when struck by a neutron. This splitting releases a large amount of energy in the form of heat, along with additional neutrons that can trigger further fission events in a chain reaction.

In a nuclear power plant, this heat is used to produce steam, which drives turbines connected to generators to produce electricity, converting the mechanical energy into electricity to be fed into the power grid. After passing through the turbine, the steam is cooled back into water (often using a separate cooling system) and recirculated to continue the process. In essence, a nuclear power plant works like a highly advanced kettle that boils water to generate electricity!

In this way, a nuclear power plant is both a complex piece of technology, and also remarkably simple in its design. Although there are many different types of nuclear reactors in the world, in essence they all work like this. None of this happened by chance – there are many stories of the scientists who pioneered nuclear technology. I would love to write a book about them to give them proper credit, but can only touch on some of them here.

The development of nuclear technology was not the result of a single breakthrough, but rather a cascade of discoveries made by brilliant scientists working across disciplines and decades. One of the earliest and most iconic figures in nuclear science was Maria Salomea Skłodowska-Curie, a Polish and naturalised-French physicist and chemist. Alongside her husband Pierre, Curie discovered the radioactive elements polonium and radium, pioneering the field of radioactivity. Her meticulous research revealed that atoms were not indivisible, as once thought, but could release immense amounts of energy. As the first person to win two Nobel Prizes, in Physics and Chemistry, Curie's work laid the foundation for all future nuclear research.

Building on these early discoveries, Enrico Fermi, an Italian physicist, made one of the most critical advances in the twentieth century: the development of the first controlled nuclear chain reaction. In 1942, Fermi and his team constructed Chicago Pile-1, the world's first nuclear reactor. This experiment demonstrated that it was possible to harness energy from the splitting of atomic nuclei in a controlled and sustained way.

The phenomenon at the core of nuclear technology, nuclear fission, was uncovered in the late 1930s by German chemist Otto Hahn, with theoretical interpretation by Austrian-Swedish physicist Lise Meitner. Hahn's experiments showed that bombarding uranium with neutrons caused it to split into lighter elements, releasing vast energy. Meitner explained the process as fission, drawing on the principles of Einstein's mass-energy equivalence. Although Hahn received the Nobel Prize in Chemistry in 1944, Meitner's essential contribution was overlooked for decades.

Natural radioactivity was first discovered in 1896, by French scientist, Henri Becquerel. Maria and Pierre Curie discovered the first artificial radioactive materials in the 1930s, which was groundbreaking for science and led to an explosion of development in industry, agriculture and medicine.

Finally, Niels Bohr, a Danish physicist renowned for his work on atomic structure and quantum theory, contributed to both the theoretical and ethical debates surrounding nuclear science. During the war, Bohr advised the Allies on nuclear fission and later promoted the peaceful use of nuclear energy.

The field of nuclear science was forged by a diverse group of scientists whose discoveries spanned disciplines and continents. Their research into the atom transformed our understanding of matter and energy and ushered in a new era of technological promise.

But long before then, in the very beginning, there was fire.

Fire was a revered gift, once worshipped by early humans. While we don't know exactly when they first harnessed fire, scientific evidence suggests that the gut of our ancestor, *Homo erectus*, began shrinking around two million years ago. This change is believed to be linked to the advent of cooking, which made food easier to digest. To cook, they must have used fire.

The first early humans to work out how to capture fire from a wild blaze and use it to cook, were undeniably pioneers too. They must have experienced a mix of reactions; some perhaps greeting their discovery with awe and reverence, with others (quite literally) getting burned and rejecting it out of fear.

Such is the power of fear. Fear of a vaccine, fear of genetically modified organisms (GMOs). Two million years later we retain the same evolutionary propensity for fear of the new or unknown, while concurrently harbouring the same immense potential for innovation and exploration. Other early humans gave the fire a chance and discovered that it kept them warm on cold nights, that it provided light on dark nights and warded off predators and flies. In the process of exploration they learned how to sharpen a flint, attach a flint to a piece of wood to create a spear, and create digging tools. This led to more effective hunting, with weapons, the ability to secure food supplies and, in turn, the formation of agriculture and civilisation. They harnessed mechanical labour from animals, wind, water; they used fire for smelting, so taking us from Stone to Bronze to Iron Age, with coal, then oil and gas, enabling the industrial revolution and lifting countless people out of subsistence living. Fire was the literal flame that ignited an explosion of science and technology. It is the story of human progress, propelled by energy.

Who was the first person to harness fire? Who thought that cooking raw hunted flesh in the fire might be a good idea? Imagine how the idea came to them, their curiosity and need to know: the first time a person sparked a flame with their hands; their surprise, awe, disbelief. The consequences of controlled fire were far-reaching. By creating calorie-dense foods through cooking, children grew stronger and their brains grew bigger. Early humans also used fire smoke to preserve meat and store it for future use.

The hearth – the communal fire – provided the vital social and cognitive focal point that played a key role in developing language, storytelling and, ultimately, what it means to be human.

Early humans were hunter-gatherers who lived in small, mobile groups. Archaeological evidence suggests that by at least 400,000 years ago, our ancestors had learned to control fire. But fire was more than a tool; it was a gathering place. The hearth created a shared social environment where people came together to cook, eat, stay safe, and interact after the sun went down.

This communal gathering was critical because social bonds form through interaction. Anthropologists argue that the hearth fostered social cohesion, encouraging co-operation and trust within groups. Unlike other animals, humans rely heavily on complex social networks. The hearth's role as a social hub gave rise to extended conversations, shared experiences, and the passing of knowledge and traditions, which were all essential for cultural development.

One of the most transformative effects of spending time around the fire was the opportunity for language to evolve. Before the hearth, communication was likely limited to basic signals and gestures, but sitting together around the fire for hours, often during the dark and cold nights, gave early humans time to exchange stories, gossip and plans. This setting would have encouraged the development of complex vocalisations and structured language, as individuals needed to describe events, teach survival skills, or share knowledge about their environment.

Psychologists and linguists suggest that storytelling was a driving force in language development. Stories around the fire helped people remember important lessons and build shared identities. According to some theories, this social interaction shaped the neurological pathways for language in the brain, allowing humans to develop the sophisticated communication systems we have today.

Additionally, controlling fire allowed early humans to migrate into colder regions, protecting them from predators and harsh environments. The ability to stay warm and active after sunset extended their waking hours, enabling more time for socialisation and mental development.

So, the hearth became the birthplace of culture. Rituals, art, music and religion all have roots in the social gatherings around fire. The storytelling that developed there transmitted not only survival skills but also shared values, beliefs and history, fostering group identity and co-operation. In essence, the hearth transformed humans from mere animals surviving in the wild into cultural beings capable of collaboration, innovation and communication. The simple energy of fire enabled us to evolve into the people we are today: a storytelling species, with language forged around a fire, now able to tell the story of fire.

By now, this story of fire may be familiar. When I first began to tell it in 2020, it was an entirely novel way to consider energy usage, and it has been retold by many other voices. Yet I continue to tell it, because I believe we cannot truly talk about the story of humankind without acknowledging this element of our journey – our deep interconnectedness with energy.

Just as some core human evolutionary tendencies have not changed over time, nor have some of our needs. Fire provided us with increased nutrition from the food we ingest, and humans still rely on energy for that purpose. But the process of obtaining energy through fossil fuels has come at a cost, because it is driving climate change and harming us through air pollution. We need a new fire: 'Fire 2.0'; energy-dense and reliable, but clean.

This leads me to an example similar to the catalyst that was fire – learning how to split the uranium atom. Nuclear energy wields immense potential. As with fire, humans learned to tame this significant discovery, and to create an energy-dense source for humanity. It was a process of trial and error in which mistakes were made, but its positive application means that our homes stay warm in winter, vital hospital equipment keeps running, and our lights stay on. All this without the lung- and heart-crippling air pollution, or global heating caused by the fire we have used since ancient times.

Humanity stands at a critical crossroads: the choices we make today about how to heat our homes and cook our meals will ripple through generations to come. The power of the atom offers a vital solution to combatting climate change, but it remains shrouded in fear and misunderstanding. Around the world, politicians uphold nuclear bans and are shutting down fully-operational nuclear power plants, so locking us into decades of more dependence on fossil fuels that continue to damage our planet. While storytellers recite countless fictional stories exaggerating the dangers of nuclear power, the real obstacle is society's persistent fear of a technological marvel born in the space age. The truth about nuclear energy's potential is sitting out there, waiting for those willing and able to look beyond the myths.

CHAPTER SEVEN

RADIATION AND RADIOPHOBIA

'Nothing in life is to be feared, it is only to be understood. Now is the time to understand more, so that we may fear less.'
Maria Salomea Skłodowska-Curie

I used to be radiophobic. Radiophobia is the excessive fear of radiation, a type of energy released by atoms in the form of electromagnetic waves or particles. On the extreme end, people with full-blown radiophobia include health patients who refuse to be X-rayed because they believe the radiation will kill them, which can even mean refusing basic dental work or radiation treatment for cancer. But, more commonly, it is a misplaced fear of everyday inventions like phones, microwaves and ovens.

The term 'radiophobia' was coined in a scientific paper in the 1920s describing people who were afraid of the new-fangled radio broadcasting, then expanded to cover a fear of newer technologies that involve human-made non-ionising radiation, including televisions, microwaves, ovens, light bulbs, loudspeakers, power lines, diagnostic medical applications such as nuclear magnetic resonance machines (NMRs), mobile phones and, more recently, 5G technology. (X-ray, CAT scans and radiotherapy do involve ionising radiation, but used in carefully controlled ways.)

Radiophobia is when a person's anxiety response is disproportionate to the type and quantity of radiation they are exposed to. One historian notes[57] that, in 1891, when electric lighting was first installed in The White House, 'President Benjamin Harrison and First Lady Caroline Harrison refused to operate the switches because they feared being shocked and left the operation of the electric lights to the domestic staff'.

Radiophobia is more common than you think, and it's such an old story, told in so many ways through apocalyptic TV viewing and books, that it may influence our feelings even when we aren't outwardly aware of it. Steve Jobs, who was once the CEO of Apple, initially refused conventional cancer treatment in favour of alternative remedies, perhaps because of the fear of radiation, but no one knows exactly why. Either way, he preferred not to expose himself to radiation treatment. Jobs was diagnosed in 2003 with a rare form of pancreatic cancer, which has a 95% chance of survival with early surgery, but he chose to forgo surgery, opting instead for acupuncture, herbal remedies and special diets.[58] Many medical experts believe that the delay reduced his chances of long-term survival. Jobs, a visionary who revolutionised technology by trusting his instincts and challenging convention, later said that he regretted his decision against early medical intervention.

People develop radiophobia through a combination of factors, including misunderstood risk perception, emotional responses and powerful storytelling, often reinforced by media, culture and activism. Unlike other daily risks, radiation is invisible, intangible, and difficult for non-scientists to understand. Because it's not something we can see, touch or feel, our brains rely on emotional cues and external narratives to make sense of it. That's not the whole story because the same argument could be made for air pollution, but fewer live in fear of it.

Over the decades, pop culture, news reports and anti-nuclear activism have helped embed a notion that radiation is mysterious, deadly and uncontrollable. Events like Chornobyl and Fukushima, although rare, are portrayed in apocalyptic terms, while the factual scientific context, such as the difference between acute and chronic exposure, is ignored and lost.

Meanwhile, we accept far greater risks from smoking, driving or air pollution, without similar fear, simply because these dangers are more familiar or normalised.

Radiophobia began with frightening new inventions, like radio transmissions and electric lighting, but different events surround the increased fear of radiation: testing nuclear weapons and distrust of government doing it; the Cold War nuclear arms race; the emotional impact of using atomic bombs over Hiroshima and Nagasaki, and falsely inflated projections of casualties after nuclear power accidents.

Pop culture has played a significant role by capitalising on our fear of radiation through various media, including films and shows, like *The China Syndrome*, *Godzilla*, *Threads* and *The Simpsons*. Many books and films involve an apocalyptic fallout scenario implying that a nuclear disaster has occurred. *Die Wolke* (the Cloud) is a German novel and film adaptation about fictional nuclear meltdown; it has been studied in German schools for decades and heavily influenced German public perception of nuclear energy. I asked German physicist Sabine Hossenfelder about what it was like as a child who had studied *Die Wolke*:

> 'It's about a girl who is about the same age I was at the time. She has a brother the same age my brother was at the time. The book is absolutely not suitable for children. Disorientating, hopeless, doesn't explain anything. I literally got nightmares from it. I've never regretted reading a book as much as I regret reading this, and that includes a lot of Stephen King. I understand now of course that it's super unrealistic. But how was I supposed to know this at age twelve? I only recovered from my nuclear anxiety when I learned how to safely handle radioactive samples in a physics lab course.

> Same way you learn to handle dangerous chemicals (I did an internship in the chemical industry). Except that radiation exposure is much easier to measure. So now I walk around and tell people to stop fussing about radioactivity. If that's what the book did to me, at least I am one data point for what it did to the German nation.'

In addition to fear driven by misunderstanding, well-funded organisations like Greenpeace continue to amplify radiophobia through misleading campaigns. One recent example is their claim that releasing treated wastewater from the Fukushima Daiichi nuclear power plant into the Pacific Ocean will lead to genetic mutations in humans, an assertion that is not supported by scientific evidence, but has nevertheless been widely reported by mainstream media.[59] In fact, three independent scientists have explained in detail[60] that additional radioactivity from the treated water will have an *infinitesimal* impact on radioactivity. They explain, 'The extra radioactivity to be added from the Fukushima water will make the most miniscule of differences… Compared with the radioactivity already present in the Pacific, the planned annual release is a literal drop in the ocean.' However, Greenpeace's messaging sparked a debate that is still raging. Sadly, this kind of fearmongering is effective and distracts from more pressing environmental challenges, as well as undermining public understanding of real versus perceived risks.

Regardless of their level of education, with constant reinforcement from media, fiction and activism that radiation is highly dangerous, it's no surprise that many people develop a creeping radiophobia. It doesn't matter if they know that the chances of being exposed to harmful levels of radiation are extremely unlikely, when they see new dystopian stories in the news that *it could change their DNA*, it changes their viewpoint.

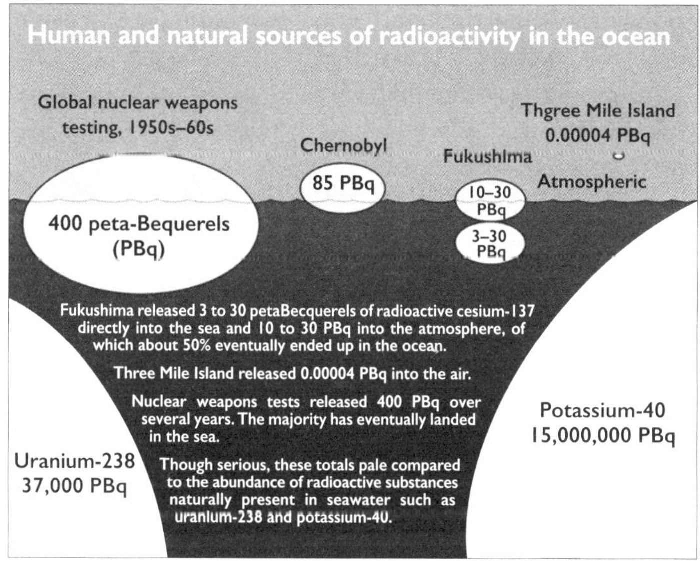

The sea is naturally radioactive

Radiation is poorly understood and frequently misrepresented by bad actors. Our level of fear often correlates directly with how little we understand the science behind it and the lack of adequate school education about energy from an early age. The more we learn about what radiation is, how it works and how it's regulated, the less frightening it becomes. We need this foundational knowledge as a buffer to when a journalist tells us that an activist organisation, with a history of spreading misinformation about a technology, has now discovered a groundbreaking new story that all the nuclear and radiation scientists and specialists in the world have somehow missed.

Radiophobia isn't just about science, but about storytelling. It thrives in a realm where fear is more compelling than facts, and where the narratives we consume don't reflect the true, manageable nature of radiation in most modern applications.

But if we look at the facts, they undermine the fear. Radiation is categorised depending on whether it is completely natural (formed from minerals in the earth), or human-made. Natural radioactive minerals are found in the ground, soil, water and our own bodies. Background radiation can come from outer space and the sun. I fact, Earth's biggest source of non-ionising radiation is the sun, in the form of visible, infrared and ultraviolet radiation (UV). For all intents and purposes, they are the same.

Radiation is all around us: in the food we eat, and even in the person who sleeps next to us at night. It's in all living things and some non-living besides. For a long time we were unaware of being surrounded by radiation, or that our bodies have naturally evolved to live with it. Our cells can stimulate DNA repair[61] to counter its impacts.

Billions of years ago, the cooled remnant was released of the first light that could ever travel freely throughout the universe; it is known as Cosmic Microwave Background (CMB) and occurred soon after the Big Bang. Without radiation, life on Earth would not exist[62] in its present form. Research[63] has found that, 'natural background radiation was, and likely continues to be, essential for evolution of life on Earth by, for example, stabilising the genome and, at the same time, allowing the necessary adaptation of organisms to environmental changes'. Radionuclides are found in foods to varying degrees, and mostly pass through our bodies when we ingest them. The radioactive form of potassium decays in our bodies, making us all slightly radioactive. Hence, bananas are above-average radioactive, because they naturally contain high levels of potassium.

Although human-made radiation is technically no different to naturally-occurring radiation, some people consider it to be more frightening. I suspect that feeling comes from naturalistic fallacy: the idea that human-made radiation is worse in some way, because we created it. But that's not true; all radiation works the same way. Not only is natural radiation found in nature, but it can also appear in large amounts, as in volcanic eruptions that can disperse large amounts of radiation. Radon isotopes are formed naturally through the radioactive decay of the uranium or thorium present in rocks and soil, and can attach to particles in the air. Radon is higher in geothermal areas[64] and leads to serious health issues including cancers. Thankfully, we can measure radon and radiation to protect people from these outcomes.

Still, I know it doesn't always *feel* that way – that something human-made, human-controlled, might be safer than its natural counterpart. And I don't blame anyone for being cautious about radiation. Fear is a deeply human response, designed to keep us safe. I understand, because I once feared radiation more than cancer. If all you've ever heard about nuclear energy is associated with accidents and radiation, it makes perfect sense to feel wary. That's not ignorance, but intelligence responding to a lack of clarity.

It took me a long time to come around to accepting what the data actually demonstrates. Like many others, I found myself more moved by dramatic stories than by abstract numbers. Again, over time, and in the wake of masses of evidence, I changed my mind. Ironically, many people who fear radiation do not have the same response to the invisible killer that is air pollution, despite the latter causing millions of deaths a year.

It is extremely unlikely that you or I will ever be exposed to high levels of radiation. For all the media focus on nuclear accidents, the reality is that modern nuclear plants are built with extensive safety measures and are subject to some of the most rigorous oversight of any industry in the world. A person living near a nuclear power plant is exposed to less additional radiation per year than someone who takes a transatlantic flight or gets a routine medical scan.

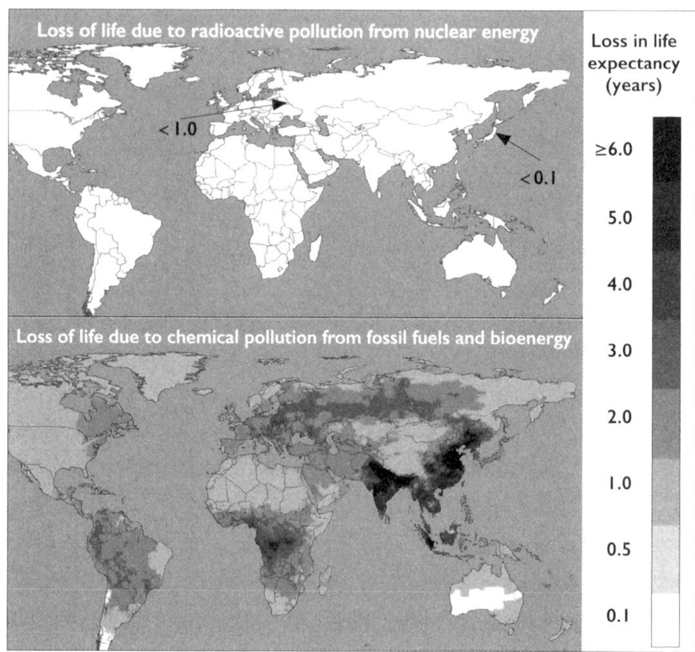

Loss of life expectancy comparison: radiation vs air pollution

That's not to dismiss people's concerns, but rather to help put those concerns into perspective. The danger, it turns out, is often more in our imaginations than in our environment.

It all comes down to risk assessment (which we are not very good at) but also numerous other factors that quickly become exhausting if we try to consider them all.

Evolutionarily speaking, we remain primed instinctively to fear things such as predators, including snakes and spiders, but we tend not to fear relatively new activities even when they are harmful to us and carry high risks. For example, the odds of dying in a car crash are 1 in 107 over a lifetime[65], a risk considerably higher than for most daily activities. Centres for Disease Control and Prevention (CDC) estimate[66] that cigarette and tobacco use kills more Americans each year than alcohol, car accidents, suicide, AIDS, homicide and illegal drugs combined, but people smoke despite knowing the risks.

Addiction hijacks the brain's reward system, primarily by overstimulating the release of dopamine, a chemical linked to pleasure, motivation and learning. Over time, repeated exposure to addictive substances or behaviours weakens natural dopamine responses, making everyday rewards feel less satisfying. This rewires the brain to prioritise the addictive substance or behaviour above all else, even in the face of negative consequences. Addiction is not a lack of willpower, but a neurological condition that reshapes how the brain evaluates risk, reward and decision-making.

So, it's not *just* about how risky something is, but myriad other factors, including how necessary or valuable we feel the invention is to us. People tend to tolerate high-risk activities if they perceive a direct, immediate benefit. Although driving comes with a significant risk of injury or death, most people accept it because it's seen as essential to daily life. Smoking, while clearly harmful, offers a short-term psychological reward, so people overlook the long-term consequences. The human brain is wired to prioritise immediate rewards over distant outcomes, a bias known as temporal discounting. Smoking delivers instant gratification: a hit of nicotine that stimulates dopamine release and temporarily reduces stress or anxiety. By contrast, health risks like lung cancer, heart disease or premature death, are abstract, delayed and seem emotionally distant.

We live in a world of many novel experiences, but from an evolutionary perspective our brains have not yet adapted to making complex risk assessments for all these activities. We have not evolved to instinctively decode the relative risks, including of air pollution, driving and climate change.

Even when we try to make in-depth assessments, considering all the factors involved with any given activity may feel complicated. We can fall prey to things like confirmation bias or reacting emotionally. For example, for many people, the idea of skydiving is terrifying even though, with only 1 injury per 1,000 tandem jumps[67], skydiving is actually safer than driving.

It's not a fair comparison though, because all risk is assessed on a scale of how safe the alternatives are and how necessary the risk is when compared to those alternatives. So, although one might criticise drivers who get behind the wheel of a car despite knowing the risks, travelling from A to B is still seen as necessary. It enables us to transport goods, travel great distances, collect our children from school, and so on, while also conserving energy, another element that humans are primed to value. Therefore, the benefits of driving outweigh its risks, which means we are willing to ignore the risk despite fatalities and accidents being so common. Meanwhile, skydiving may be a fun occasional activity but it doesn't have the benefits for our lifestyles that driving has, so although the risk of skydiving is statistically lower, not being perceived as necessary makes it feel greater.

It is easier to overcome a fear when we are exposed to it yet don't experience negative impacts. For example, I was afraid of spiders as a child, although after years of encountering spiders and not coming to any harm I have rationally made peace with spiders and overcome my anxiety around them. Equally, the more we undertake an activity like driving, the more normalised it becomes and the less we fear it. This explains why most people who live by nuclear power stations tend not be afraid of them. And people who work in them see them as necessary because they understand the technology so well. On the other hand, it's difficult to overcome a fear of flying when we do it so seldom, even though logically we all know that it's statistically safer than driving.

Some of my family in India. My mother is second from the left, and I'm the youngest child on the right.

In the BBC studio during my interview on The Andrew Neil Show.

Leading an anti-McDonald's protest in London, 2004 – one of my early acts of environmental activism.

Taking part in an anti-nuclear protest in Plymouth, 2010.

With French nuclear advocates, Voices for Nuclear, after speaking at their rally in Lyon, 2021.

At COP26 in Glasgow, I recreated AOC's iconic Met Gala dress – this time carrying the message 'Build more nuclear'.

During a projector action in France, French advocates Voices for Nuclear displayed two messages: 'Fessenheim, we miss you' and 'Nuclear saves lives'. Photo credit: Voices for Nuclear.

Emergency Reactor held stalls and rallies across the UK. This event took place in Bristol, 2021.

Reactivists dropped banners in London, 2021.

At an Emergency Reactor protest outside Greenpeace HQ in Islington, London, 2022.

We delivered charcoal to represent coal to Greenpeace HQ in London, 2022.

Modern-day treehugging. This storage cask from Sizewell B represents 4 billion KWh of zero carbon electricity – enough to run nearly 1.3 million homes for a year.

We got the government's attention: projector action in London on the Department for Business, Energy and Industrial Strategy, 2020.

Theatre piece at COP26 with US advocates from Generation Atomic. 'Married by scientific consensus, but coal tried to stop the nuclear-renewables wedding.' Glasgow, 2021.

Rally for Sizewell C on the beach with Emergency Reactor in Suffolk, 2021. Decommissioned Sizewell A, with B visible in the background.

Let's be frank: in very high doses, radiation is harmful. But that is true of many things, including drinking too much water,[68] taking too much paracetamol, or eating too many apples. When making any assessment, looking at the quantity or dose we are exposed to is important. As the saying goes: *the dose makes the poison.*

A sievert (Sv) is a unit that measures the biological effect of radiation on human tissue. It tells us how much energy from radiation has been absorbed by the body and how harmful that dose is likely to be. Because one sievert represents a very large dose, smaller units like millisieverts (mSv) and microsieverts (µSv) are more commonly used for everyday exposure levels. One sievert equals 1,000 millisieverts, and one millisievert equals 1,000 µSv, helping us better quantify and understand radiation doses in practical terms.

The dose threshold for acute radiation syndrome (ARS) begins at around 1 sievert, which is equal to 1,000 mSv. At this level, symptoms like nausea, fatigue and lowered white blood cell counts can occur, especially if the dose is received over a short period. For long-term health risks, particularly cancer, studies of populations exposed to radiation, such as atomic bomb survivors and radiotherapy patients, have shown a statistically significant increase in cancer risk at cumulative doses above approximately 100 mSv. Below this level, the risk is much smaller and harder to detect against background health data, though regulatory bodies often apply the precautionary principle.

For the average person, reaching radiation doses anywhere near 100 mSv, let alone 1,000 mSv, is extraordinarily unlikely. Natural background radiation from the Earth and cosmic rays expose most people to 2–3 mSv per year, depending on where they live. Even a single chest CT scan delivers only about 6–7 mSv, and a transatlantic flight gives you around 0.03–0.05 mSv due to higher cosmic radiation at cruising altitude. In nuclear power plant operations, strict safety measures ensure that both workers and the public are exposed to only tiny fractions of these thresholds. Outside of rare nuclear accidents or intensive medical treatment, like radiotherapy, exposure to harmful levels of radiation is almost non-existent in everyday life.

Let's put these measurements into everyday context. Some common foods are naturally radioactive, though at extremely low levels. For instance, a single banana's potassium content exposes you to about 0.01 μSv of radiation. Similarly, eating Brazil nuts also exposes you to roughly 0.01 μSv. These tiny doses highlight how natural background radiation is all around us, and perfectly safe.

The banana-equivalent dose is often used to illustrate examples like this; miniscule amounts of radiation exposure that do no harm.

Banana Equivalent Dose

Bananas are a natural source of radioactive isotopes.

Eating one banana = 1 BED = 0.1 μSv = 0.01 mrem

No of bananas	Equivalent exposure
100,000,000	Fatal dose (death within 2 wks)
20,000,000	Typical targeted dose used in radiotherapy (one session)
70,000	Chest CT scan
20,000	Mammogram (single exposure)
20–1,000	Chest X-ray
700	Living in a stone, brick or concrete building for one year
400	Flight from London to New York
100	Average daily background dose
50	Dental X-ray
1–100	Yearly dose from living near a nuclear power station

Meanwhile, global background radiation that we are exposed to every day, is around 2.4 millisieverts. Near where I live, in the South West of England, we have above-average exposure to radiation because the granite rocks below us are rich in naturally-occurring uranium and thorium. People living south of me, in Cornwall,[69] are exposed to 6.9 millisieverts a year, due to minute amounts of radon in the uranium that is present in all Earth materials. About 50% of the heat given off by the Earth is generated[70] by the radioactive decay of elements such as uranium and thorium, as in many other areas worldwide. Also, people living at high elevations, like in Denver, USA, are exposed to higher levels of cosmic radiation from space.

None of this radiation exposure harms people in any of these places.

On the high end of the scale, a person who smokes twenty cigarettes a day is exposing their lungs[71] to between 0.4 mSv up to a whopping 53 mSv every year due to the polonium in the tobacco they are smoking. The reason these numbers vary so wildly is because scientists aren't sure how much radiation is actually absorbed by a smoker's lungs. This brings us back to difficulties with making risk assessments, since there's a high risk involved with smoking: tobacco smoke contains a radioactive chemical element called polonium-210, which emits alpha-radiation, which can seriously damage DNA. When exposed to polonium-210, human skin usually acts as a barrier, but the same cannot be said for when it is inhaled. One study[72] estimated that smoking a pack-and-a-half of cigarettes every day exposes a smoker to a dose of radiation equivalent to 300 chest X-rays a year.

Meanwhile, astronauts are exposed[73] to between 50 to 2,000 mSv in space because it is difficult to fully protect them from natural cosmic radiation. On Earth, we are protected from the full impact of solar and cosmic radiation by the Earth's atmosphere and magnetic fields.

There are 440 nuclear reactors currently operating in the

world, which shows how rare these accidents are. Accidents and health consequences from fossil fuels are considerably more common and deadly, but because they don't have the same amount of compelling storytelling behind them, they haven't permeated public consciousness like the idea of a radioactive disaster.

It's interesting that many of us – and I once included myself in this – believe that nuclear power plants pose the biggest risk of radiation, because to generate the same amount of electricity, a coal power plant gives off at least ten times more radiation than a nuclear power plant.[74] Yet, the stories that stay in our mind are those of meltdowns and nuclear war, which solidifies the link between nuclear energy and danger.

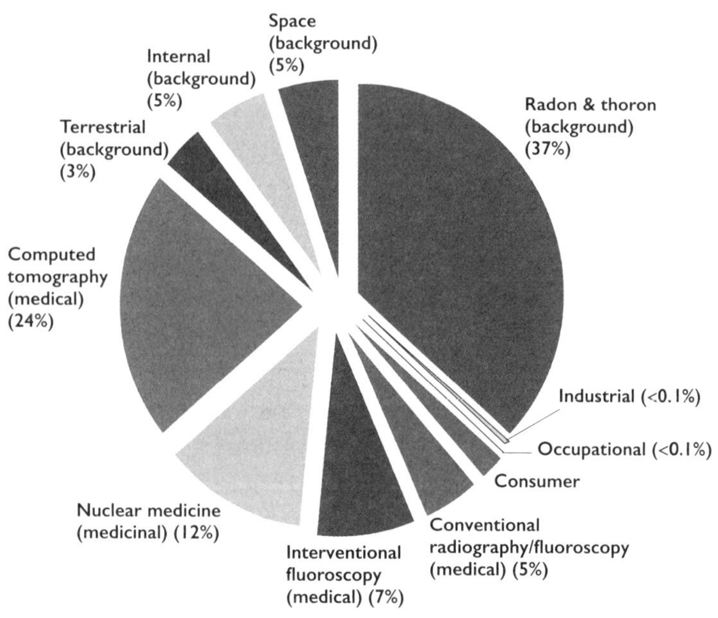

Sources of radiation exposure

When I first spoke out publicly about nuclear energy in 2020, I was acutely aware that it had a perception problem: its benefits are abstract or invisible (reliable baseload power, low emissions), while the perceived risks of radiation and meltdowns are ubiquitous, dramatic and emotionally charged in popular culture. Because many people don't connect electricity to survival or quality of life in a visceral way, they view nuclear energy as optional rather than essential.

But consider this: no one talks, or protests, about nuclear medicine, which is also a life-saving technology that produces nuclear waste. Nuclear medicine is a medical speciality that uses radioactive tracers (radiopharmaceuticals) to assess bodily functions and to diagnose and treat disease. To name just a few examples: it helps determine whether organs are functioning normally and whether the blood supply to the heart is adequate, it detects cancers at an early stage, checks whether the heart can pump blood adequately, identifies abnormal brain lesions (without invasive surgery), assesses whether kidneys are functioning normally, and ascertains lung function and bone density. In short, it helps to keep us healthy.

Medical sources, particularly diagnostic X-rays, are one of the most significant human-made sources of radiation. If you have a broken or fractured bone, you want to know exactly what's damaged and where, so that you can get the correct treatment. And nuclear medicine saves many lives; over 20 million Americans[75] benefit each year from nuclear medicine procedures used to diagnose and treat a wide variety of diseases. So is the resistance to nuclear energy just about the technology, or about something else?

I have come to the conclusion that energy itself has a PR problem. Most people feel that it's bad to use too much energy, that they should use less, and that it's to blame for many of our problems. This has been the dominant messaging from the traditional environmental movement for decades; and it has convinced world leaders and shaped policy worldwide.

That's why it becomes critical to explain that energy is life-sustaining. Without electricity there's no healthcare, no clean water, no food refrigeration, no internet, no modern economy. If more people understood that *energy is the foundation of everything we value in society*, there would be a deeper appreciation for clean, reliable sources like nuclear energy, and more proportionate feelings towards elements like radiation.

The main thing to understand about radiation is that the only time you're likely to be exposed to significant amounts is by choice – probably as part of a life-saving medical procedure.

CHAPTER EIGHT

WHAT ABOUT THE WASTE?

'All of the nuclear waste from all of the energy that the US has produced from nuclear energy since the 1970s, which is about 20% of all of our electricity for 50 years, has not generated enough spent fuel to fill up a football field above 10 metres high.'

US Department of Energy

'What about the waste?' is the question I get asked the most. The idea of 'nuclear waste' used to fill me with dread. I believed it was a green, gloopy, corrosive liquid as depicted in *The Simpsons*. I thought it was dumped in rivers, contaminated the environment with radiation and gave people cancer. I feared waste as if it *was* cancer. This fear was closely linked with my fear of radiation. I was eighteen years old and believed that nuclear energy was dangerous, the nuclear industry was evil and renewables were the only way to save the planet. So why did I change my mind?

I learned that nuclear energy is essential for decarbonisation, that nuclear power stations have the smallest land footprint of all energy sources per unit of electricity they produce, and that the energy they produce is clean[76] and reliable. However, one thing continued to concern me: nuclear waste.

I had been told by countless friends, and green organisations of which I was an active member, that huge amounts of nuclear waste could not be disposed of safely and it was leaking into our world's precious environment. It was years before I realised that I had been misinformed by people who were themselves, misinformed.

Ironically, I feared nuclear waste as if it was cancer when coal, the most reliable alternative to nuclear energy, is actually causing not only cancer,[77] but respiratory issues and other serious health problems.[78] Yet nuclear medicine saves millions of lives every year. But there's good news, because waste is nothing to be feared, and there also isn't very much of it.

Nuclear wastes are classified according to how radioactive they are. Low-level and medium-level wastes include things like used protective clothing, tools, wiping cloths and other disposable items that have been used on site. They are not particularly hazardous, but because of the nuclear industry's high safety standards they are treated with rigorous caution; so much so that steel and concrete from nuclear plants can only be recycled as scrap if their radiation levels are a thousand times lower than the same materials from the oil and gas industry.[79]

High-level radioactive waste consists of irradiated, or spent, nuclear reactor fuel. It consists of small solid fuel pellets in long metal tubes called rods. A structured group of fuel rods is called a 'fuel assembly'. I was wrong: there is no scary green liquid. The waste is not even liquid.

What I had been told about leaking waste was nothing to do with nuclear energy. The leaky vats I was afraid of were from weapons-related reactors, such as the Hanford Nuclear Site in Washington State, where the US Department of Defense and Department of Energy produced plutonium for use in the atomic weapon programme. Some of these contaminants leaked into the land and water, including into the Columbia River.

I was even confused about how much nuclear waste there is in the world. It turns out that there isn't very much of it. All the high-level nuclear waste produced in the world would fit into a single football field to a height of approximately ten yards.[80]

The amount of high-level waste produced during nuclear energy production is also small: a typical large reactor produces about 25–30 tonnes of used fuel per year. Worldwide, 97% of waste produced by the nuclear power industry is classified as low- or medium-level waste.[81] In France, where fuel is reprocessed, just 0.2% of all radioactive waste by volume is classified as high-level.

When a fuel assembly is removed from a reactor because it can no longer provide useful power it is extremely hot, both thermally and radioactively. It is immediately transferred to a cooling pool where it is stored alongside other used fuel as it cools down for two to ten years and becomes less radioactive. After this, it can be moved to a dry storage cask.

Once I realised how wrong I'd been about what nuclear waste is, I was determined to learn how it is managed. This required a lot more work as I had to understand how waste is managed by other industries, and how they manage fossil fuels and renewable production, to give a fair comparison.

We live in an industrialised civilisation where waste is a by-product of our everyday lifestyles. Many wastes are hazardous, including lead, mercury, arsenic, cadmium, chromium, chlorine, hydrofluoric acid, cyanide, asbestos, dioxins, many other carcinogens, clinical wastes and various pathogens. They are managed carefully, for the most part. But nuclear waste is, arguably, managed the most carefully of all.

I held a misconception about nuclear waste and radioactivity, in which I believed it would stay highly radioactive for many hundreds of thousands of years. I learned that, unlike other industrial toxic wastes, radioactivity – the principal hazard associated with high-level waste – diminishes significantly with time. This is known as radioactive decay. The amount of time it takes for the radioactivity of radioactive material to decrease to half its original level is called the radioactive half-life.

The decay of heat and radioactivity over time means that

after only forty years, the radioactivity of used fuel has decreased to about one-thousandth of its level compared to when it was unloaded. Less than 1% is radioactive for 10,000 years. This portion can be easily isolated and shielded to protect humans and wildlife.

When spent fuel must be transported, say to reprocessing plants, it is carried in dry storage casks. The containers are extremely strong and have been rigorously tested by being dropped from heights, having railway locomotives crash into them[82] and missiles flown at them.[83]

When the fuel assembly has cooled down, it is transferred to a stainless steel and concrete dry storage cask designed to contain fuel assemblies for around a century. The casks shield radiation, meaning you can stand next to one, unexposed to radiation from the waste inside. These casks are usually stored at the power station. I'm hugging a dry storage cask in the photo taken at Sizewell B in Suffolk.

Fuel rods contain 90% of the potential energy from the uranium in them. In some countries, such as France, Japan and Russia, used fuel is reprocessed to extract usable fuel to make new fuel assemblies. Through this method of recycling, up to 96% of the reusable material in spent fuel can be recovered. France's national policy of recycling spent fuel[84] has meant that it needs 17% less natural uranium to operate its plants than it would without recycling. There is currently enough energy in the United States' nuclear waste to power the entire country for a hundred years with clean energy.[85]

These days, it's a no-brainer to me to say we should go green and recycle our nuclear waste. Ultimately, something is only waste if you waste it.

Once it has been recycled multiple times, the waste can be returned to the earth, by which I mean it can be buried.

The international scientific consensus is that deep geological repositories are the most effective approach to permanently disposing of high-level radioactive waste. Deep geological repositories are already used for long-term disposal of waste containing arsenic, cyanide, mercury and other toxic chemicals. Finland has created the world's first deep geological repository to bury high-level nuclear waste 450 metres below ground level, and Sweden too is constructing their first underground repository.

When I was twenty-four, I attended an anti-nuclear protest because of what we environmentalists believed was a silent killer – radiation from waste. But radiation hasn't harmed anywhere near as many people as fossil fuels. I now realise that we live with real silent killers every day: air pollution from burning fossil fuels, and the slow march of climate change.

It may seem like overkill to focus so much on nuclear waste, but it's important to put things into perspective. To truly understand the risks and benefits, we need to compare like with like; specifically, nuclear waste versus the waste generated by other forms of energy production. Only by making such comparisons can we appreciate in context the relative scale and manageability of nuclear waste. As you read this, waste from fossil fuels is being stored in the Earth's atmosphere, and we breathe it in every day. Air pollution from fossil fuels causes millions of deaths a year.[86] Oddly, this deadly invisible killer doesn't instil fear in people the way the idea of nuclear waste does.

Waste from renewable energy technologies is rarely recycled, largely because doing so is technically challenging and economically unviable. As someone who championed renewables for years, I was shocked to discover that the solar panels I had long promoted as a green solution are often discarded in landfills,[87] where they leach toxic chemicals into the soil,[88] and often in developing countries where environmental safeguards for the population are weaker. I also learned that most batteries are not recyclable and that wind turbine blades pose similar disposal issues.[89] This isn't to say we should abandon renewable energy, but rather to highlight a crucial truth – *all* forms of energy production come with environmental trade-offs. No solution is perfect, or without environmental and human impact. I was twenty-nine when I confronted this reality, and I was heartbroken. As renewables scaled up, I began to question how we would truly phase out fossil fuels while still meeting rising energy demands and protecting our planet. It was time for me to seriously reconsider nuclear energy.

What about the waste generated from mining? The truth is, every energy source, and nearly every aspect of modern life, depends on mining. Your phone, the chair you're sitting on, the food packaging you just threw away, even the table you ate at; all required the extraction of raw materials. Energy is no exception. The difference lies in how much we need to extract to keep things running. This is where nuclear energy stands out. It's powered by uranium, one of the most energy-dense materials on Earth. This means that a small amount of uranium can produce a massive amount of energy, causing far less mining and far less environmental disruption compared to other sources. If, like me, you care about efficiency, minimising harm and reducing waste, this matters. Nuclear energy offers a way to meet our growing power demands while keeping our environmental footprint as small as possible.[90]

Ultimately, we should be recycling nuclear waste. First, there is currently enough energy in the stored nuclear waste in the United States to power the entire country for 100 years.[91] That's 100 years of clean energy sitting in storage casks, doing nothing.

Thanks to its high-energy density, very little nuclear fuel is required to produce large amounts of electricity, especially compared to other energy sources. This means that the resulting waste is very small; on average, the waste from a reactor supplying a person's electricity needs for a year would be about the size of a brick. Roughly 5 grams of this is high-level waste, which is about the same weight as a sheet of paper. Most of the material in used fuel can be recycled; in fact as much as 97% of it. In some countries, like France, Japan and Switzerland, used fuel is already reprocessed, but other countries, most notably the US, treat used nuclear fuel as waste and leave it sitting in storage casks, doing nothing.

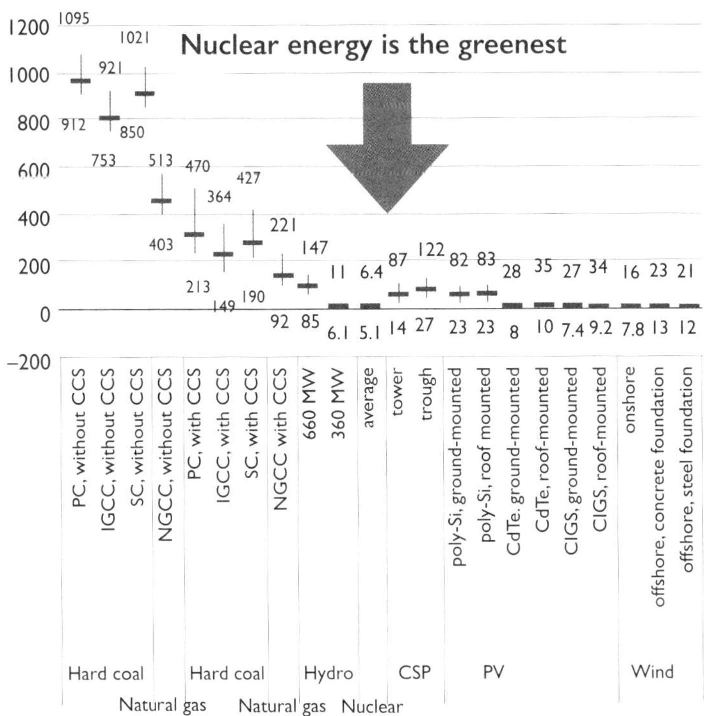

Lifecycle greenhouse gas emission ranges for the assessed technologies

You may be surprised to learn that the technology to turn nuclear waste into energy, known as a nuclear fast or breeder reactor, is not a new technology. It has existed for decades and, here's another surprise, *it was developed in the US in 1951*. The Experimental Breeder Reactor 1 (EBR-1) was a nuclear reactor that could use nuclear waste as fuel. It was developed by a US government research facility called the Argonne National Laboratory. This place was a hub of scientific discovery – nearly all operating commercial nuclear power plants worldwide have roots in research that took place at Argonne.

A breeder reactor is a nuclear reactor that generates more fissile material than it consumes. Using them would enable us to power the planet for billions of years without much need for freshly mined uranium. The EBR-1 was the world's first breeder reactor and one of the world's first electricity-generating nuclear power plants. It initially produced enough electricity to illuminate four 200-watt light bulbs and later generated sufficient electricity to power its building.

The vast majority of the world's nuclear power plants – roughly 70% of the global reactor fleet – are light-water nuclear reactors, but almost any reactor design can be tweaked to become a breeder. The light-water reactor has evolved into the fast reactor concept, where the fission chain reaction is sustained by fast neutrons.

The reasons for recycling waste are obvious (Japan cites energy security, conservation and reduced waste production), but the most compelling reason that most people aren't aware of is that recycling spent fuel reduces its radioactivity and, therefore, the time it requires in storage. We currently have to store spent fuel for tens of thousands of years, but by reprocessing the waste, the radioactivity and corresponding need for storage can be reduced to *just 200 years*.

If you're afraid of radiation, you should love the idea of recycling spent fuel, as it's the best way to reduce its radioactivity. Whether or not you like the idea of new nuclear power plants, there is existing nuclear waste stored in facilities around the world, and there is a proven method for significantly reducing its radioactivity. This method resolves both the need to store the casks for longer periods of time and also provides clean energy without the need to mine new resources. It is, surely, any environmentalist's dream come true.

Finally, a note on the final resting place for waste. For that very end by-product, minimal though it may be, the scientific consensus is that the best way to manage it is to permanently store it in a deep geological repository (DGR). This is a specially designed underground facility for the long-term storage and disposal of high-level radioactive waste, such as spent nuclear fuel. The idea is to isolate dangerous materials deep within stable layers of rock, hundreds of metres below the surface, where they are unlikely to be disturbed by natural events or human activity. Waste is sealed in protective containers, often surrounded by additional barriers like clay or concrete, and then placed in tunnels or chambers carved into the rock. This multi-barrier system is designed to store radioactive substances for tens of thousands to hundreds of thousands of years.

Some nuclear advocates argue that DGRs are excessive, suggesting they reinforce the idea that nuclear waste is inherently dangerous, which only strengthens existing fears about spent fuel. On the other hand, many experts within the scientific community believe that long-lived, high-level radioactive waste demands a secure storage solution capable of safeguarding the material for thousands of years. Both sides make valid points.

I understand why some nuclear advocates are concerned that discussing long-lived nuclear waste in this way might make the issue seem greater than it is. However, I also believe we need to be cautious about letting anti-nuclear activists shape decision-making to the point where we avoid discussing critical issues like permanent long-term storage. Many other industries store waste underground without controversy. As is often the case, it's not the dramatic problem anti-nuclear groups make it out to be.

Finland has already constructed the world's first long-term repository for radioactive waste: Onkalo, which translates to 'pit' or 'cavity' in Finnish, is 450 metres underground. Onkalo is essentially a vast network of tunnels designed to isolate spent nuclear fuel. The facility features robotic vehicles to transport the waste and a ventilation system to maintain air quality. I spoke with a Finnish MP who told me that, in future and for a price, Finland could allow other countries to store waste in their repository.

Sweden is following in Finland's footsteps and is in the process of constructing the world's second deep geological repository for nuclear waste; the Forsmark Deep Geological Repository. The site is located near the Forsmark nuclear power plant, and the project is expected to be operational sometime in the 2030s.

Storing nuclear waste in casks for decades is fine, and we certainly shouldn't bury it while it can still be recycled, as that would be wasteful. However, once the spent fuel has been fully used, it should be returned from whence it came, to deep underground. This means openly discussing permanent waste storage rather than avoiding it. After all, many industries manage waste produce and bury it in DGRs without sparking controversy; they are already an accepted method of long-term disposal of waste containing arsenic, cyanide, mercury and other toxic chemicals.

I made a TikTok video about DGRs that went viral and the most asked question in the comments was on how future civilisations can be made aware of this waste. It's a fair question. My answer is that language doesn't disappear overnight. It changes and evolves over time, slowly, so there's no reason for modern civilisations not to update instructions for DGRs – for all waste, not just nuclear waste – to ensure that it's in legible modern language rather than the equivalent of a Shakespearean script to future generations.

At the moment, scientists do the following: durable markers, such as massive stone or concrete structures, are placed at the site to last thousands of years and convey a sense of danger. Universal symbols, pictograms and inscriptions in multiple languages aim to communicate risk even if modern languages are lost. The goal is to make the site intuitively uninviting and clearly hazardous, combining engineering, design and human psychology to protect people far into the future. Ultimately, there's no reason that we can't update this information as we go along. There's also no reason that technologically capable future civilisations wouldn't be able to decipher past language and symbols or, at the very least, ascertain what's underground by measuring the radiation themselves. We can already do this today, but I would be surprised to learn that in tens of thousands of years humankind hasn't found a better way of measuring the same thing.

CHAPTER NINE

ASSESSING AND BYPASSING RISK

'We live in a society exquisitely dependent on science and technology, in which hardly anyone knows anything about science and technology.'
Carl Sagan

It's oddly freeing: words that once conjured terror – 'nuclear waste', 'Chornobyl', 'Fukushima' – no longer fill me with terror. If you have worries about nuclear technology, I hope sharing this information can be freeing for you, too.

Data from Oxford University offers perspective on all the nuclear meltdowns that have taken place, and shows that fossil fuels are still significantly more harmful than nuclear energy.

Since those meltdowns mentioned in Chapter 4 occurred, nuclear power plants have been made safer than ever. The RBMK reactors that are still in operation have been upgraded. Authorities like the IAEA work to monitor and ensure the safe, secure and peaceful uses of nuclear science and technology, even basing themselves[92] at the Zaporizhzhya nuclear power plant in Ukraine and holding talks with Russian officials to ensure the safety of the plant (where, despite a fire outside the site and being taken over by Russians, no one has been harmed because of the power plant). Nevertheless, there was a lot of media hype and scaremongering around Zaporizhzhya. Meanwhile, fossil fuels continue to kill hundreds of us daily. (In 2022 the Ukrainian government reaffirmed its commitment to nuclear energy.)

Stories of radiation scare us because we've watched them play out on screen so many times in the world of fiction, because

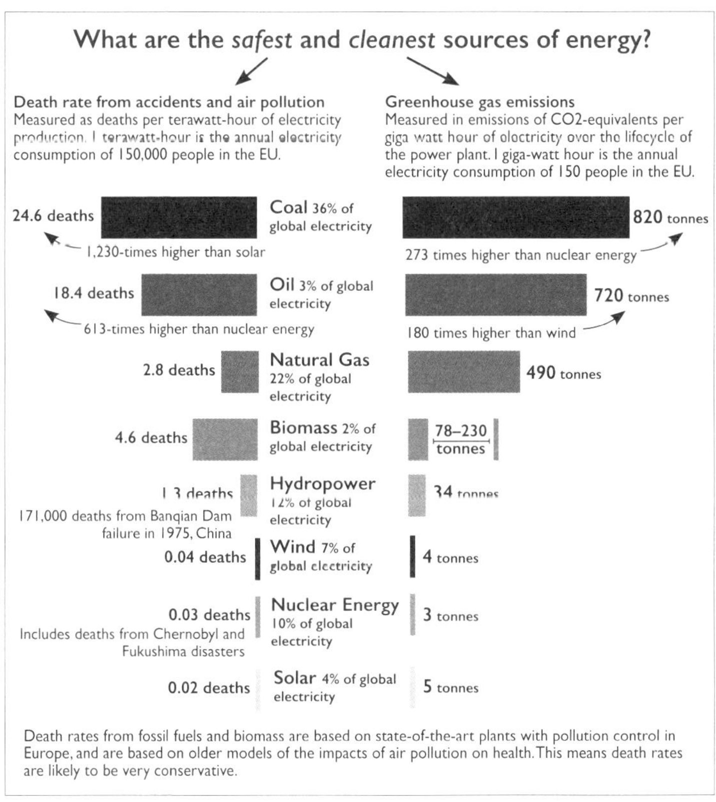

The safest and cleanest sources of energy

people have spread misinformation about disasters, and because we conflate nuclear meltdowns with the fallout from atomic bombs. The outcome of such storytelling is that we are frightened of the wrong thing.

Nuclear waste is one of those things.

Another is the argument made about nuclear meltdowns; that they displace populations. Yet, in total, more people have been displaced by other forms of energy generation, specifically coal mining. For example, in the Lusatia region of Germany alone, 30,000 people were displaced by coal mining,[93] and more than 130 villages were deliberately destroyed. Globally, coal mining displaces millions.[94]

Large dams built for hydropower, irrigation, water storage or flood control, have also led to the involuntary displacement of millions[95] over the last century. China's Three Gorges Dam displaced over 1.2 million people[96] and, in India, over 16.4 million people have been displaced due to hydropower development projects. In 2000, the World Commission on Dams estimated that from 1950–2000 the number displaced directly by large dams was between 40–80 million people. This figure has increased in recent years.

For a phenomenal example of the greater danger of traditional energy sources, consider the Jharia coalfield fires, which have been burning underground in India for over a hundred years.[97] The first underground fire there was ignited by unknown factors in 1916, and the fires now cover over a hundred square miles of Jharkhand State in north-eastern India. Various attempts at putting them out have failed. It is estimated that 37 million tons of coal have been consumed by the fires since their start, the total emissions from which are unknown. Look them up. They look truly apocalyptic, with people's homes sinking into the fires below ground as the earth they stand on melts beneath them.

Again, oddly, no disaster films have been made about these fires, which are still raging.

There has been sadly little research on the health impacts of these fires, but the toxins they release make people sick, sometimes fatally. People suffer from health issues including respiratory problems, carbon monoxide poisoning, stroke, lung cancer, pulmonary heart disease and chronic obstructive pulmonary disease. As of 2007, more than 400,000 people who live in Jharia are living on land that is in danger of subsidence, it could sink beneath their feet at any time.[98] Coalfield fires are more common than people realise, and Jharia is just one of thousands of fires around the world that are burning underground. For example, over 200 coal fires in Pennsylvania in the US have contributed to making it one of the leading acid rain producers in the States.[99]

By any comparison, nuclear power plants are ridiculously safe. The powerful earthquake that recently hit Turkey and Syria killed over 34,000 people. The majority of these deaths were a direct result of poorly made buildings collapsing onto people. Yet we were barely informed that particular earthquake did not damage the Turkish Akkuyu nuclear power plant at all.[100]

The 2010 Deepwater Horizon oil rig explosion, on the other hand, caused the largest marine oil spill in history. A surge of natural gas blasted through a concrete core, and petroleum that leaked from the well before it was sealed, formed a slick extending more than 57,500 square miles over the Gulf of Mexico. Two hundred million gallons of oil and two million gallons of chemical dispersants leaked into the sea, with a devastating impact on the environment. An estimated million coastal birds died[101] and the discovery of dead corals near the spill indicates that it harmed marine life in the deep ocean.[102] A study[103] found heart abnormalities in fish embryos exposed to Deepwater Horizon oil. By 2013, over 650 dolphins had been found[104] stranded in the oil spill area, a four-fold increase over the historical average in that area. Elevated numbers of people reported abnormal symptoms including respiratory problems, blood in urine, rectal bleeding, seizures and miscarriages.[105]

Oil spills are sadly regular occurrences. Remember also the Amoco Cadiz oil spill, the Exxon Valdez oil spill, the Persian

Gulf War oil spill, and many others.[106] On the other hand, radiation can easily be measured, treated, contained or shielded, and is simply nowhere near as dangerous or harmful.

To illustrate this point, I think we should normalise nuclear waste by putting it in public places that allow people to see and understand it. In the Netherlands, the Central Organisation For Radioactive Waste (COVRA) stores the country's high-level waste in casks in a public museum and art gallery that also hosts exhibitions. It alters our perspective, by showing how well managed nuclear waste actually is.

When undertaking risk assessment for any technology, we should consider this: what are the risks of *not* building it? Failing to develop safer energy sources, advanced medical technologies, or transportation infrastructure doesn't mean zero risk; it means higher energy costs, slower progress, environmental degradation, and lost opportunities to improve lives. Every decision to halt innovation carries its own chain of consequences, often far more damaging than the hypothetical worst-case scenarios we fear. A truly responsible assessment must weigh both sides: the risks of building, and learning from, technology, and the risks of remaining stagnant.

Anti-nuclear sentiment is largely a Western phenomenon, and that is not a coincidence. The people who have most benefitted from abundant energy – largely from fossil fuels – have been the most vocal against it because they have never had to endure energy scarcity. In countries like India, families like mine and millions like them, are eager for access to reliable and safe energy. For them, nuclear energy isn't an abstract debate about cost or safety, but a matter of achieving cleaner air, better health and a better future for their communities. Who else wouldn't want that for their children?

They do not have to unlearn the fear of clean energy,

because they have not been told the story of nuclear fear for decades. Instead, they live with the realities that energy scarcity brings: dark nights without lighting, homes stifling in relentless heat without air conditioning, and the daily struggle to power the essentials of modern life. For them, the dangers of energy scarcity are far more immediate and tangible than the abstract fears that dominate Western debates about nuclear energy.

This hints at a dark truth: that across the West we have been taught, and learned, to fear nuclear energy.

Now I am older and have two children. I have hugged dry fuel casks and understand that each cask represents four billion kilowatt-hours of zero-carbon electricity; enough to run nearly 1.3 million homes for a year. I tell this to my children. Every day when I wake up, I am grateful to be able to provide my family with a warm, safe home, because I know that billions of people around the world do not yet share this privilege.

I believe in an energy-abundant future, one informed by science rather than fear of things we don't understand. After a decade of being afraid, I now recognise that nuclear energy is essential for the future and that nuclear waste is a positive part of that future.

Realising all this about waste, and then seeing how it is managed and stored, marked a turning point for me; from being someone who protested nuclear energy, into someone who now champions it globally. Knowledge is power, and releasing the fear of non-understanding has been liberating. I hope it will be for you too. Because the true waste is the amount of time we spend worrying about spent fuel instead of celebrating what this clean, reliable source of energy offers us.

CHAPTER TEN

CHALLENGING NEGATIVITY BIAS

'The world is indeed full of peril, and in it there are many dark places; but still there is much that is fair, and though in all lands love is now mingled with grief, it grows perhaps the greater.'

J.R.R. Tolkien, *The Lord of the Rings*

When I first joined XR, the movement was alive with hope and urgency, but within a year its tone had shifted – from energy and possibility to a drumbeat of doom. This shift often happens in social movements. What starts with energy, optimism and the promise of change do, over time, become dominated by fear and negativity, sometimes even silencing the voices that originally inspired it.

Within XR, there was ongoing, intense debate over the direction the movement should take. Some members argued that only extreme action would capture attention and convey the urgency of the climate crisis – and later these individuals formed a splinter group, Just Stop Oil. Others believed that inspiring and energising people with solutions and achievable goals would have a more lasting impact. Over time, however, the narratives focused on doom and catastrophe began to dominate. Fear became the primary lens through which the climate *crisis* was communicated, and the original ethos of hope, empowerment and practical action was gradually overshadowed.

In 2023, I spoke on a panel with the director Oliver Stone at an opening for his new documentary. Afterwards, he told me that when he first heard me speak, beginning with the words 'poverty is energy poverty' on stage, a light switched in his mind. It was something he always sensed, he said, but he had never heard expressed so clearly, and he wished I'd been in his film. Coming from the man who directed *Platoon* and *Natural Born Killers*, I was honoured to hear this from such a masterful storyteller.

Again, this highlights the divide between the scientists who discover and invent new technologies, and the storytellers who shape society. Although, notably, despite being a masterful storyteller, Stone chose to create a scientific documentary rather than a fictional film. This decision is significant, because popular culture remains starved of positive fictional narratives about nuclear energy. While documentaries can inform us, fictional stories have the power to challenge deeply held beliefs, normalise new ideas, and shape public perception, which nuclear technology still very much needs.

A useful tool for challenging melodramatic beliefs and ideologies that we don't always realise are there, such as overpopulation of the planet, is Gapminder, a non-profit organisation that was founded by Hans Rosling, his son Ola Rosling, and daughter-in-law Anna Rosling Rönnlund. It is a data visualisation platform designed to produce global statistics on health, wealth and development, accessible and engaging; it helps people understand complex data through interactive charts and animations.

On its website, Gapminder invites users to take a multiple-choice quiz[107] testing how well we understand the world. Famously, the team discovered that humans scored worse than monkeys – who were pressing buttons at random – on most of the questions. Thankfully, there is still time to stop being a monkey.

In XR, the narrative was often centred on the idea that the human race is 'doomed' or that the planet 'only has four or ten years left'. While such urgent claims have captured public

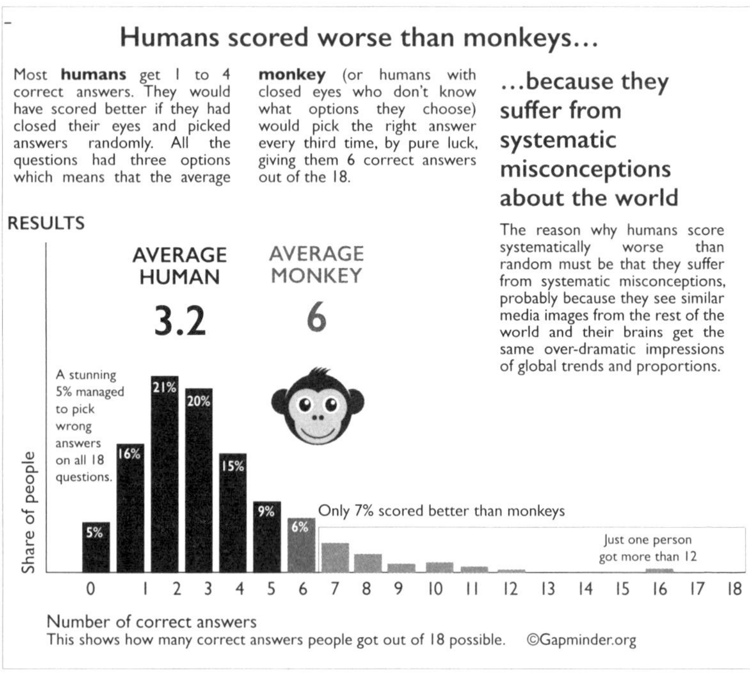

attention, it's important to recognise that these timelines are speculative and unsupported by scientific consensus. There is no established tipping point or irreversible deadline for the planet's survival (no, predictions from isolated rogue scientists don't qualify for consideration). Ultimately, Earth will endure with or without us, even if the climate undergoes severe changes, becoming more like that of Venus.

People once feared that the ozone layer was beyond repair. It was, justifiably, the environmental panic of its time, with many activists warning that the Earth's end was near. Scientists rallied to communicate the risks, and world leaders responded. Almost all nations united to adopt the 1987 Montreal Protocol on Substances that Deplete the Ozone Layer,[108] and around

99% of harmful chemicals were phased out. Remarkably, the protective ozone layer began to heal itself. The Antarctic ozone hole is now projected to close by the 2060s, while other regions are expected to recover to pre-1980s levels even sooner. Thanks to these efforts, an estimated two million lives are saved each year from skin cancer. Additionally, studies[109] show that without the Montreal Protocol, the lack of increased carbon storage in plants, vegetation and soil would likely have added 0.5–1.0°C to global warming.

The evidence demonstrates that we are fully capable of solving the world's greatest challenges. The real folly lies in treating apocalyptic predictions as wiser or more insightful than optimistic, solution-focused perspectives. In truth, pessimism is simply the easier path, fuelled by our natural negativity bias. Yet, numerous factors demonstrate that in many key areas, humanity is moving in a positive direction.

That's not how most of us feel about the state of the world, and I can understand why. For a while, I too stopped opening my social media apps because the news has been continuously bleak. The details are depressing and horribly graphic. It is a double-edged sword: we have more easy access to information than ever before, but that means the bad news reaches us in detail as we sit down to dinner with our families. We simply aren't wired to take in this much information, let alone hear about bad things happening all over the world that we can do nothing about. The bad news can feel all-encompassing.

Negativity bias is a psychological phenomenon in which humans tend to focus more keenly on negative information, experiences or emotions than on positive ones. This bias evolved from the survival mechanism of paying closer attention to threats and dangers, which helped our ancestors avoid harm. However, in today's world, negativity bias can distort our perception of reality, making problems seem larger or more insurmountable than they actually are. It leads people to give greater weight to bad news, fears or failures, often overshadowing positive progress and achievements. Negative bias fuels pessimism and fear, making it harder to appreciate solutions and to hope, even when clear evidence shows things to be improving.

The traditional environmental movement has often tapped into this negativity bias, persuading millions that the planet is doomed, that climate change is an unstoppable catastrophe, and that humanity is to blame. I believed it once, and am here to tell you that none of these beliefs hold up under scrutiny. While serious challenges remain, and humans have made mistakes, the narrative of inevitable doom overlooks the substantial progress humanity has made, and continues to make, towards solving environmental problems.

We *are* capable of decoupling growth from emissions. We are capable of improving air quality to make us healthier and makes our children taller, smarter and less violent. We are capable of tackling climate change as well as eradicating poverty. We are capable of meeting net zero targets.

None of this is headline-grabbing news, so no one is likely to shout it from the rooftops (although I have tried). But humans have repeatedly proven our ability to solve enormous problems, including those we ourselves have created and many we have not. We have been doing so for years. Betting everything on failure is a disservice to humankind's potential and to how unimaginable goals can be achieved by working together. After all, no difficulty is ever solved by fixating solely on the problem. It's time to focus on implementing evidence-based solutions, to build the cleaner, healthier world we all aspire to live in.

Actually, I am approaching this the wrong way. I should be explaining how we are already achieving incredible things and doing extraordinarily well. Because we are doing better than ever before, on almost every metric. In recent years, scientists, innovators and policymakers have achieved major wins that offer tangible hope for global health, the environment and technology.

To begin with, many aspects of human life have improved worldwide over recent decades. The past thirty years have seen immense improvements in the quality of life for much of humanity. Extreme poverty has fallen by nearly two-thirds,[110] from 1.9 billion to around 650 million. Life expectancy has risen in most of the world, as have literacy levels and access to education. Infant mortality rates have dramatically fallen.

Data shows that under 1% of people in high-income countries now live in extreme poverty; around 70% of all people now have easy access to safe drinking water; more children have access to education than ever before; and child mortality rates are consistently on the decline.

Advances in science and technology are improving lives in countless areas. One of the most significant developments is the rollout of highly effective malaria vaccines in Africa. With the World Health Organization (WHO) approving the R21/Matrix-M vaccine alongside the earlier RTSS, countries are launching large-scale immunisation programmes.[111] This is an historic shift against a disease that kills hundreds of thousands of children each year, and experts believe it could save millions of lives over the coming decades.

In the fight against pollution, researchers have engineered plastic-eating enzymes that can break down polyethylene terephthalate (PET) bottles in hours rather than centuries.[112] These enzymes are already being tested in pilot recycling facilities, offering a path towards reducing landfill waste and ocean plastic pollution by turning used plastics back into raw materials.

Another success story is that we have now passed 'peak farmland', which means that we now produce more food using less land. Global agricultural land use reached its highest point in the early 2000s, and since then it has been shrinking. Across the globe, former farmland is reverting to grasslands, forests and shrublands, allowing wild animals to reclaim areas they once roamed freely, and thanks to irrigation, improved seeds, fertilisers and pesticides, we've spared around 1.8 billion hectares of land from being farmed – which is three times the size of Spain.

Life-changing medical breakthroughs are myriad. A baby born today with a health condition that was considered incurable just twenty years ago now has a far greater chance of living a full, healthy and normal life, thanks to remarkable advances in medical research, technology and treatment.

In London, doctors have restored partial sight to four children born blind due to a rare genetic condition called Leber congenital amaurosis (LCA), by using a pioneering gene therapy.[113] The treatment involved injecting healthy copies of the faulty *AIPL1* gene directly into the retina, allowing the eye's cells to produce a crucial protein needed for vision. As a result, children who had never seen clearly before were able to recognise shapes, play with toys, and even begin reading. The breakthrough, carried out at Moorfields Eye Hospital, marks the first time this specific therapy has been successfully used in humans and offers hope to other patients with similar inherited retinal diseases.

In a major breakthrough for type 1 diabetes treatment, researchers have successfully used stem cell-derived islet cell transplants to restore natural insulin production in patients. In a new clinical trial, a patient who had been dependent on daily insulin injections for decades was able to produce sufficient insulin on his own after receiving lab-grown islet cells, which replace the pancreatic cells destroyed by the immune system in type 1 diabetes.[114] This meant he no longer required external insulin for blood sugar control. The results represent a potential shift towards a functional cure. I could go on.

These are recent news stories at the time of writing. They may not be as prevalent on your TV screens as conflicts and instances of violence, but they show that while challenges remain, progress is real. The future is not just about the problems we face, but about the solutions we create, and these breakthroughs prove that human ingenuity is very much alive.

While violence and war still exist, these upward trends of improving life for people and reducing suffering are promising. It's easy to be pessimistic and it often feels wrong to celebrate the wins when there are still challenges ahead – thanks, negativity bias! – but let's take a moment to applaud how far humankind has come. Our lives are filled with miracles of science and technology thanks to 300 years of the scientific method. Western civilisation has mostly overcome superstitious thinking, barbarism and totalitarianism. We have fought for democracy, against slavery, towards equality, and we have achieved many of these aims the world over. Humanity is, in many ways, thriving. The Earth isn't doing too badly either – yes we need to tackle air pollution, but we also saved the ozone layer, and there is plenty of other good news too.

Childhood vaccinations are estimated to save 3 to 5 million lives every year worldwide, according to the World Health Organization. These vaccines protect against deadly diseases like measles, polio, diphtheria, tetanus and whooping cough, which once claimed millions of young lives annually. Since 1990, global efforts to expand immunisation have contributed to a dramatic decline in child mortality.

Maternal mortality rates have also improved dramatically over the past century, due to advances in medicine, sanitation and access to skilled healthcare. In the early twentieth century, maternal death was a tragically common outcome of childbirth in many parts of the world, even in wealthy countries. Today, a woman giving birth is far more likely to survive the experience, with global maternal mortality having declined by nearly 40% since 2000 alone. Improved access to prenatal care, antibiotics, emergency obstetric services and trained birth attendants, has saved countless lives. While disparities still exist, particularly in low-income countries, the overall trend is positive, reflecting how far we've come in protecting mothers and babies through evidence-based care.

Estimates suggest that nuclear energy has saved around 1.8 million lives globally by displacing coal and other fossil fuels. This figure is based on avoided air pollution-related deaths due to nuclear power generating electricity without emitting harmful particulates, nitrogen oxides or sulphur dioxide – common byproducts of coal combustion. The same study also estimates that if nuclear power were replaced by fossil fuels instead of being expanded or maintained, millions more deaths could occur in the coming decades due to increased air pollution and greenhouse gas emissions.

But these are just numbers, and we know numbers can only take us so far.

We have become so comfortable with our progress, so safe relative to previous decades, and so healthy, that we only see the bad aspects of our lifestyles rather than all the good we have achieved. This is the privilege of having so much, and it's worth mentioning the Tocqueville Effect.

Named after the French thinker, Alexis de Tocqueville, it describes the paradox that social unrest often grows not when conditions are at their worst, but when they begin to improve. As people see progress, whether in political rights, economic opportunities, or social equality, their expectations rise faster than reality can meet them. This gap between what people experience and what they believe they deserve can fuel frustration and even revolution. Tocqueville observed this in the lead-up to the French Revolution, where some reforms had been made, but the remaining inequalities became more visible and intolerable. In modern contexts, the Tocqueville Effect helps explain why movements for change often intensify after partial progress, as progress itself sharpens awareness of what is still lacking.

And so, it was easy for tales of overpopulation, of humans as a virus and harmful to nature, to override the story of our successes. Some of us began to refuse vaccines and protest 5G masts. Governments lost their way in maintaining progress energy-wise; instead of pushing for more clean energy we went backwards, aiming for less. A society that understands the need for progress and energy would never enter into an energy crisis. We have forgotten what fuels us, how integral industrialisation is to our lifestyles today, and how fortunate we are to have these privileges. We have lost sight of the bigger picture – and the foundation it is built on.

We in the West are among the world's richest people, capable of solving immense challenges, and first and foremost of those is climate change. If we don't truly understand the state of the world and the challenges we face, how can we hope to make things better?

Few of us believe this story of life improving, partly because of cognitive biases, and partly because they don't *feel* true because the successes are largely *invisible*. It's hard to appreciate the lives saved through phasing out coal, or increasing access to childhood vaccinations, or the babies that made it to childhood who would not have survived a century ago. It's much easier to tell the story of a life lost through violence or disease, but when those issues are being tackled, how do we spread the good news without relying only on numbers? That's why I often contrast my life in Britain with the way my relatives in rural India live. Once, all of humankind lived in poverty like them.

It is an immense achievement that we have found ways to improve life for so many people, and it speaks volumes about humankind and our priorities and motivations. People aren't inherently as destructive or selfish as popular narratives often suggest. Time and again, research and real-world examples show that when given the right tools, knowledge and incentives, humans are capable of cooperation, compassion and ingenuity. Recognising this doesn't ignore the problems we face – it highlights that we have the power to solve them when fear and cynicism stop holding us back.

If you are unsure about, or afraid of nuclear energy, and you have made it this far in the book, I want to commend you for taking the time to allow your beliefs to be challenged. I know from experience that it isn't an easy thing to do.

CHAPTER ELEVEN

ADDRESSING THE APPEAL TO NATURE FALLACY

'I've come up with a set of rules that describe our reactions to technologies:

1. Anything that is in the world when you're born is normal and ordinary and is just a natural part of the way the world works.

2. Anything that's invented between when you're fifteen and thirty-five is new and exciting and revolutionary and you can probably get a career in it.

3. Anything invented after you're thirty-five is against the natural order of things.'

Douglas Adams

If you're reading this you likely lead an energy-rich lifestyle. I certainly do.

I come from a working-class background. Raised in a council house by parents who worked in factories; I grew up in the inner city where being street smart was a necessity. Having seen my parents escape energy poverty first hand, I never quite felt comfortable with the middle-class environmentalist narrative that using energy is inherently bad. For my family, gaining access to reliable energy and, alongside it, appliances like washing machines, didn't feel like a luxury but liberation. Owning material items represented the hard-won fruits of decades of sacrifice and struggle. A family car wasn't just convenience but freedom; the ability to participate fully in society. And since much of our extended family lived in a region without electricity or internet, the idea of giving up flying didn't feel like a noble gesture, but like further isolation from a home my parents had to leave behind.

My own core values and life experience clashed with environmental perspectives on almost every issue, but at the same time, I was young, deeply concerned about climate change, and still working out how to articulate my views.

As an active member of the UK Green Party in 2015, the then leader, Natalie Bennett, alongside prominent environmentalist, Bill McKibben, endorsed my book on low-carbon living.[115] However, during a panel discussion with Bennett some years later, she remarked that people in the UK 'do not have a high quality of life'. While she was probably referring to the mental health crisis, her comment echoed a common assumption that people who 'live on the land' in close-knit communities must naturally enjoy better mental health because of their lifestyles.

Natalie was lauding the 'naturalistic' fallacy, a logical fallacy that occurs when people assume that something is good because it is perceived as 'natural', while it is bad because it's 'human-made' and therefore 'unnatural' or 'artificial'.

The very basic premise of the appeal to nature fallacy is that anything natural cannot be wrong, as if naturalness is itself a kind of authority. The fundamental problem with this romanticised argument is that it's impossible to define what 'natural' means. After all, aren't humans natural beings? If so, isn't anything we do to ensure our survival therefore natural? Instead, we typically think of the tools we make as forms of technology and lump them into groups of what we feel is natural versus what isn't. With the nature fallacy, it's easy to pick apart the argument once you dig a little deeper than the surface vibes, but still, the idea persists across populations worldwide.

Worse, there are countless examples of authority figures

making decisions based on the appeal to nature fallacy. Consider the prevalence of organic or 'natural' food and clothing, on which people spend large amounts of money because they believe the marketing that states it's better for them. There's also 'chemical-free' food (which, if it were true, would be a miracle, since *everything* is made of chemicals), and misinformation about processed foods, which people fall for because it intuitively *feels* like good sense.

Then there are 'alternative medicines' like homeopathy, which, in England, was funded by the National Health Service until 2017; add reiki, naturopathy, herbal medicine, and, in the US, there's 'energy healing' which involves using magnets for 'magnetic therapy'. You might laugh, but a poll found that 14% believed this to be 'very scientific', with 54% of participants calling it 'sort of scientific'. So, maybe you won't laugh. Perhaps you, too, feel magnets can alleviate pain and help your body to heal. If so, you are, of course, entitled to your beliefs, but without empirical evidence to show their efficacy, shouldn't we worry that, in some instances, people are being misled and suffering because they are being misled by false solutions? The same can be applied to the way we address environmental challenges as well.

Stories are powerful and they're hard to rewrite. Pseudoscience does often inform policy, influence what gets built, funded and what gets banned. Multiple studies into these examples of pseudo-medicine have found[116] that they lack biological plausibility, testability, repeatability and evidence of effectiveness, but the placebo effect exists, and people use alternative remedies despite all evidence to the contrary, feeling that they work because they are more 'natural' than proven alternatives.

In a world increasingly shaped by scientific progress, public resistance to well-established scientific consensus can seem baffling. From opposition to genetically modified organisms and nuclear energy, to vaccine hesitancy and outright climate change denial, these narratives span a wide range of emotional intensity and ideological commitment.

To meaningfully engage with these views, and counter harmful misinformation, it helps to place them on a spectrum that ranges from light-hearted observation to melodrama.

At the mildest end of the spectrum lies light-hearted scepticism or cultural preference. Here, people may reject certain technologies or innovations based on personal values, lifestyle choices or cultural norms rather than concrete scientific objections. For example, while buying organic foods or using homeopathy alongside approved medicine can influence market trends, they rarely cause direct harm, unless they discourage the use of effective medical or environmental interventions. Believing in astrology and reading horoscopes is also on the light-hearted end of the spectrum.

Moving along the spectrum, we find emotional concern or misunderstanding. This category includes people who are genuinely worried about safety but have encountered conflicting information or poor scientific communication. For example, a parent might express concern about vaccine side effects or nuclear accidents, without fully subscribing to misinformation. These views are shaped by fear, personal experience and a lack of trust in institutions. Importantly, individuals in this group may be open to changing their minds when presented with respectful dialogue and clear evidence.

A more intense and potentially damaging form of narrative emerges with ideologically-driven rejection. Here, opposition to science is no longer based on misunderstanding but rather on moral or political beliefs. Anti-GMO sentiment may be framed as resistance to 'corporate control of agriculture', while anti-nuclear positions may invoke fears of ecological disaster or industry manipulation. These narratives are often fuelled by mistrust in authority (The Man), and their proponents tend to cherry-pick data that supports their worldview while dismissing opposing evidence. This level of rejection begins to influence public policy and has resulted in blocking solutions to urgent issues like climate change and food insecurity.

At the extreme end of the spectrum lies melodrama and conspiracy thinking. These narratives are emotionally charged, highly theatrical, and often apocalyptic. They are not simply sceptical of science, they turn it into an active enemy. Claims that gene-edited crops cause farmer suicides, or that nuclear energy is a death machine disguised as energy policy, fall into this category. The logic is often circular, immune to correction, and bolstered by charismatic influencers or online echo chambers. These narratives pose a real threat to public health and scientific progress, particularly when they spread rapidly via social media.

Understanding the different levels of resistance helps us respond more effectively. Those expressing mild scepticism often benefit from respectful conversation and trustworthy information. Those entrenched in ideology may require more nuanced engagement, such as exposing internal contradictions in their arguments. And with conspiracy-level narratives, the focus may need to shift towards protecting vulnerable populations from harm, rather than trying to change the minds of deeply committed adherents.

Contrary to popular belief, falling for the nature fallacy isn't related to how much education a person receives – for example, obtaining a degree does not necessarily mean that you have been taught media literacy and can identify misinformation. Appeal to nature fallacy is endemic, crossing cultures, political lines, class systems and religions. For example, the highly educated King Charles III is the Patron of an organisation that promotes homeopathy. He uses a homeopath and also advocates for organic farming.

Then there is the sheer number of celebrities who market expensive products on the basis that they are 'natural'. To name a few famous examples, we have actress Gwyneth Paltrow's promotion[117] of vaginal steaming, jade eggs and cleansing enemas; actor Woody Harrelson's campaign[118] that 5G networks cause respiratory infections and Covid; and activist Russell Brand's 'magical amulet'[119] which he claims can protect users from 'evil energies' and retails for a whopping $239.99. It's easy to make fun of uterus cleansers and magic amulets, but the sad fact is that fearmongering works, and countless victims fall for it. Note that these are all exceptional storytellers, able to captivate their audiences and use stories to sell their wares.

Perhaps you might say that they are harmless, and we can debate whether inserting an egg into your vagina is harmful or not, but, notably, in this instance, Paltrow's company, Goop, faced a lawsuit and settlement for making false advertising claims. And the basic premise of these products is to use them instead of regular medicine. I've even heard some anti-medicine advocates argue that illnesses that can't be treated with alternative medicine are 'nature's way' of taking care of population size – which brings us full circle back to misanthropy again. But make no mistake, when used to replace conventional medicine, these alternatives extend suffering and claim lives. And if people are actively being frightened into using those products and away from actual medicine, that should be a cause for concern too.

If you consider it for long enough, you'll probably find that you have some beliefs based on the nature fallacy. Here are a few examples that you may have heard and believed: our food system is broken; pesticides and fertilisers are making us sick; living on the land is better than modern civilisation; chemicals are harmful; organic foods and farming are better for human health and the environment; radiation is unnatural and dangerous;

GMOs are causing mutations in humans; vaccines create autism; 5G masts are causing cancer and Covid; sugar is making children hyperactive, and so on. Anti-nuclear arguments also have the nature fallacy at their core.

So let's address them. Food productivity has never been better thanks to advances in technology; subsistence farming means returning to poverty and most people don't want to live that way; pesticides and fertilisers enable us to feed millions of people; everything consists of chemicals; organic farming is not better for the environment or human health; radiation is natural, surrounds us, and humans are radioactive; virtually no food is natural as humans have changed it in some way at some point; gene-editing saves lives; vaccines do not cause autism – and there's nothing wrong with being autistic – and even if it were true it would still be preferable to dying from measles, mumps, rubella, polio, whooping cough or diphtheria; recent studies suggest that sugar may actually have a calming effect because it produces serotonin. Nuclear reactors are the most natural form of energy with the most significant benefit for nature; but making this statement is still seen as controversial and biased despite masses of data to back it up.

Occasionally, an appeal to nature perspective that most people find absurd rears its head publicly and enables people to marvel at it together. Recently, several posts about drinking 'raw water' from natural sources went viral, because those who posted believed the water to be 'natural' due to how clear it is. However, this water is dangerous to drink and rife with bacteria, viruses and parasites that result in waterborne diseases including cryptosporidiosis or giardiasis. It looks stunning but it is deadly. Since most people do know this, the internet was unforgiving towards the people promoting 'raw water', but before casting the first stone we should perhaps take a closer look at our own blind spots.

Another example of a widely accepted myth is degrowth theory, which has gained popularity in recent years. Degrowth advocates for deliberately reducing economic activity in order to lessen humanity's environmental impact, arguing that endless growth is incompatible with a sustainable planet. While it may sound appealing in theory, the movement often romanticises poverty and overlooks the essential role that growth has played globally in improving health, education and living standards. Despite these concerns, degrowth has gained significant traction, including receiving endorsement from the European Parliament,[120] illustrating how quickly abstract ideas can be embraced without fully considering their real-world consequences.

I used to be a hippy, in every sense of the word. I spent my early twenties climbing trees to stop them being cut down, living in communes, helping to raise chickens and trying to grow enough food to feed a community. So, I can say with some authority that hippies have typically believed that humans have it wrong, that civilisation and development are evil (The Man), and that 'living on the land' with minimal technology is the only way to make peace with nature. But, as I say, this clashed with my own life experience, and the irony of having to handwash clothes in a commune was not lost on me at the time; my parents had travelled 5,000 miles and left behind everything they knew, to avoid having to do this sort of back-breaking labour every day of their lives back home.

Just as laborious, time-consuming labour seems beautiful to people who consider it to be more natural than using a washing machine, they still have to wear clothes and use toothbrushes that they didn't make from scratch. Interestingly, they also tend to find eternal flames sacred. These are natural flames that are kept burning continuously by a steady supply of fuel that are found all over the world. In almost every country where one exists, it is considered sacred by the local population. In some cultures, eternal flames serve as a symbol of remembrance, honour or eternal life. Again, this is perhaps, because fire is one of humanity's oldest and most essential discoveries, symbolising warmth, survival and the spark of civilisation itself – our first real tool to transform the world. Fire tells a story that is hard for us to resist.

But the same people who visit eternal flames with a sense of wonder, also dislike the idea of burning natural gas, coal or methane, which is what's happening with these flames. In Mount Wingen in New South Wales, Australia, an eternal flame caused by a smouldering coal seam underneath the surface has been burning for over 6,000 years.[121] Coal seam fires are estimated to contribute 3% of the world's annual carbon dioxide emissions, releasing mercury and other pollutants into the atmosphere. Even though it's a coal fire, the idea prevails that it must be pure because nature created it, and 'sustainability' sites like Treehugger call it a 'fascinating phenomenon' that 'holds spiritual significance in some cultures and religions'. Apparently, when nature is damaging the planet, that's okay.

Nature fallacy may seem problematic on closer inspection, but it's incredibly persuasive. It's almost as if we are wired to believe in it. As with all logical fallacies, it's possible to combat the appeal to nature fallacy by examining and challenging our ideas and beliefs. Conversations with peers regarding the danger of nature fallacy can also be effective when approached in a relatively indirect, non-confrontational manner. I'm of the opinion that logical fallacies and other critical thinking skills should be taught in schools to provide buffers for people who are more susceptible to them. Had these skills been taught when nuclear energy was first implemented, there may not have been such a successful backlash against nuclear energy during the hippy golden era.

It strikes me as odd that 'wokeism' has faced such a backlash, when arguably the most damaging forces to our planet have been certain factions of anti-science environmentalism. While the former is largely cultural and symbolic, the latter has directly hindered the adoption of technologies, like nuclear energy and gene-editing, that could genuinely mitigate climate change, protect ecosystems and save lives. Who are they? Greenpeace, Friends of the Earth, and certain factions within Extinction Rebellion, GeneWatch UK, The World Wildlife Fund and Sierra Club, to name a few.

Worse still, these anti-science factions haven't just slowed progress at home – they've actively influenced policy and public opinion abroad, blocking or persuading other countries from developing life-changing technologies.

If you donate to any of these organisations, I urge you to reconsider.

My family of rice farmers in India aspire to have what we enjoy in the West – reliable energy, development, including access to healthcare, education, lighting, air conditioning, refrigeration to preserve food and medicines, and more. For them, life tied to the land is not idyllic but filled with hardship: real risks of rabies from dog bites, deaths from snake bites, and relentless toil in sweltering 40°C heat. What might seem, from a distance, like a simple rural existence, is in reality a daily struggle for survival. They do not consider this to be a high quality of life.

CHAPTER TWELVE

HOW WE CAN CREATE A HIGH-ENERGY, LOW-CARBON UTOPIA

'We are a remarkable species. We're capable of so much.'
Captain Kathryn Janeway, *Star Trek Voyager*

Poverty was once the norm. A quarter of babies died in their first year of life.[122] In 1980, around 40% of the world's population lived in extreme poverty.[123] Today only 10% of people do.[124]

Much of this is thanks to fossil fuels. The burning of wood, then coal, gas and oil, enabled us to prosper. This had a significant impact on the environment and climate. For most people who live in a building of any kind, that building was built on what was once a forest, grassland or desert. So was most of the world. To live as a human being means that we constantly make trade-offs with the environment. It has been this way since the dawn of humankind.

Trade-offs may sound obvious, but despite the positive impact on humankind of harnessing energy, environmental groups around the world continue to protest energy – all sources of it – and politicians have followed their lead by taking clean, reliable energy out of the equation.

It's easy to become downhearted by the bad news associated with energy production, but we have seen that accidents must be considered in context, and our energy choices need to be informed by data, not scary stories.

Ultimately, the consequences of lacking development are more significant than all these accidents put together. We need energy to survive. Without it, we would be trapped in poverty. Billions of people around the world still are. Poverty is, essentially, energy poverty. To continue to thrive, warm our homes, light our streets, run hospitals and schools and feed billions of people, we need more reliable, clean energy.

I spent over a decade trying to convince people to consume less and reduce their carbon footprints, but the best behavioural science in the world has not found a way to make this happen enough for it to make a real difference in terms of climate change. Most people choose convenience over other options: this is an easily-understandable evolutionary trait. We can and should insulate buildings, fund public transport and take other measures to avoid energy wastage, but even if we all do this the world over, it will not stop climate change or mitigate our growing need for more energy.

We take energy for granted. In a way, this is a good thing as it means we have moved past having to worry about the lights working after dark and being able to wash our clothes without hard manual labour. However, we still feel guilty for using energy, thanks to traditional environmentalism that has convinced us humans are a curse on the planet and we once lived in simpler, therefore better, times with fewer creature comforts and limited technology. Reducing personal energy use has become a widespread focus ⊠ that globally informs policy.

A recent example of a *Washington Post* journalist proposing we should bathe in cold water to reduce energy usage[125] demonstrates the ongoing glorification of energy scarcity by the most energy-rich individuals. The writer argues that heating water harms the planet as it 'gobbles energy', but neglects to cover the fact that being alive also gobbles energy. There is no way around that, discounting the most ardent Malthusian who believes in population reduction by force. The technologies we access in energy-rich nations, such as home heating or hot water and soap to kill germs, are not superfluous to our needs but essential for basic hygiene and healthy lifestyles.

Virtue signalling about reducing one's own carbon footprint may make journalists feel good, but the impact of lifestyle choices like bathing in cold water is frankly infinitesimal given the scale of the problem. In order to live well, humans use a lot of energy. Overall, hyperfocusing on individual footprints distracts us from the effective large-scale solutions that are crucial for improving life on this planet.

We can invest in renewables, but we also need to be honest about their shortcomings. Wind and solar still need baseload power and wherever nuclear energy has been phased out it has historically been replaced by coal or gas.[126] Hydropower is more reliable, but dams are geographically restricted and carry significant risk.

The root problem is not growth itself but that we have not always balanced human needs with protecting biodiversity and the health of our planet. Notably, wildlife thrives around existing nuclear power plants, as is shown in the exclusion zones around Chornobyl and Fukushima.

In fact, the exclusion zone around Chornobyl is now a flourishing nature reserve, home to thriving populations of animals including wolves, eagles, deer, lynx, boar, beavers, elk, bears and the endangered Przewalski's horse.[127] Wildlife is also prospering around the Fukushima power plant.[128]

Ecologist Jim Beasley, who studies wildlife in contaminated areas of Chornobyl and Fukushima, found that populations of animals have been healthily increasing in both regions, despite the high contamination. 'I've never seen an animal with an outward visual deformity from radiation,' he asserts. Even in the most contaminated areas, none of the seventeen species that Beasley and his team documented were affected by radiation levels.[129] The Chornobyl exclusion zone is now home to European bison and previously rare species that were introduced to the area after the accident to help with their conservation. Przewalski's horse is one of the most endangered animals on Earth, previously so rare that only a few decades ago it existed nowhere in the wild. A population now thrives in Chornobyl.

Beasley comments that 'this image or perception of Chornobyl [as a barren wasteland] that I had might not be a reality… Our simple presence and use of a landscape can in some cases be more detrimental to the long-term survival of a species than a nuclear accident.'

By contrast, the areas that are used for extracting fossil fuels are ravaged in the process and do not always recover. In light of this, to turn our backs on nuclear energy is to turn our backs on the health and life of the planet.

We need to choose energy sources that do least harm to the environment and human health, and to do this, we need to consider our options fairly and view them in context. Clean energy is essential for replacing fossil fuels that significantly harm us and our planet.

When we panic and have knee jerk reactions towards nuclear energy, we ignore the consequences of mining and burning coal, the millions of deaths from air pollution, the oil spills and gas-related accidents, the millions of displacements, and the impact of burning fossil fuels on our climate. Through fear, we continue to make decisions in darkness. When we call for nuclear power plant closures around the world based on misplaced narratives, we invite literal darkness. Meanwhile, in Jharia and other regions where coalfield fires rage, the ground is literally melting beneath people's feet. That's what we should really be having meltdowns over.

We will always need more energy, not less. The shift towards electrification and data centres is already demonstrating this. Consider also what may be next: around the corner could be solutions that save millions of lives and propel humankind forward, but if we embrace energy scarcity now, we will be unable to power these future discoveries and technologies. Energy scarcity has taken centre stage this decade because people worry about climate change and want to achieve net zero goals, so they have increasingly embraced degrowth. But data shows that we can have it all if we embrace nuclear. Cleaner energy without unrealistic sacrifices. If not nuclear, then what? With current technology, we can have energy-rich lifestyles without causing planetary destruction.

Air conditioning, AI and water desalination plants are energy intensive but necessary means for solving world problems, as long as we create the evergreen energy needed to power them. Many of us take clean drinking water for granted. While some people complain about this complacency, I actually think it's positive because it means we don't know what it's like to experience deadly dehydration. That said, we should think beyond short-term needs and avoid becoming complacent about important issues like the growing need for clean water. So how do we spend energy?

1. We need energy for: desalination

Water stress attracts media attention but is seldom presented with an accompanying solution, as if it is inevitable and inescapable. With news that the Three Mile Island nuclear power plant is due to reopen to power Microsoft data centres, people are catching on to the idea that we need more energy to power AI; but what about the need for clean water? According to the United Nations, 2.2 billion people globally currently lack access to safe drinking water, and freshwater is in short supply for an increasing proportion of humanity. Although roughly 71% of the Earth is covered in water, only 2.5% of it can be consumed or used for crops.

In addition, the supply of available freshwater is in decline due to droughts becoming more common,[130] and the 1.1 billion people who inhabit the world's driest regions are most at risk of experiencing water scarcity. When people think of water stress, they might imagine death from dehydration, but water scarcity also impacts food production and causes diseases to spread. In regions of the world affected by water stress, yield gap closure depends on sufficient irrigation, and some researchers[131] estimate that water demand for crops is likely to increase by at least 146% around the middle of this century. Water scarcity also leads to sanitation problems that result in waterborne illnesses, such as typhus and cholera.

As the poet Coleridge once lamented, water is everywhere but we usually can't drink it. That was the story in the eighteenth and nineteenth centuries but, thankfully, humans are innovative creatures and now there are various methods of converting saline water into fresh water. We already know that in order to turn seawater into drinking water we have to remove the dissolved salt. This requires distillation, which is the process of boiling seawater, capturing the steam, and condensing it back into water. One of the most effective methods of cleaning up vast quantities of water doesn't get as much attention – desalination.

The concept of desalination is not new. In 350 BCE, the Greek philosopher Aristotle[132] observed that 'salt water, when it turns into vapour, becomes sweet, and the vapour does not form salt water again when it condenses'. Desalination is one of the oldest forms of water treatment, and was used on ships in ancient times to convert seawater into drinking water. There are several methods that achieve the same outcome, including reverse osmosis and thermal desalination.

The first patent for desalination was granted in seventeenth-century England, and Britain built the first seawater distillation plant in Yemen back when the country was a British colony. The technology was neglected for many years, but revived in 1930 by an unlikely candidate – the island of Aruba.

Desalination plants typically last thirty to sixty years or more, although their lifespan depends on the technology used. Plants only take a few years to build. For parts of the world where fresh water is not readily available, desalination is a thirst-quenching, often life-saving, technology. There are currently almost 20,000 desalination plants in operation around the world; some are small production plants serving residential or private industrial areas, while others are large and supply an entire population or industry. The potential for desalination plants to deliver water is tantamount to the invention of the aqueducts that enabled Romans to settle lands far from water sources. Over half of all desalination plants are in the Middle East,[133] where the largest are located in the United Arab Emirates, Saudi Arabia and Israel.

In 1998, the Middle East was hit by the worst drought it had experienced in 900 years, and which lasted over a decade. It hit farmers heavily in countries like Israel, Jordan, Turkey, Lebanon and Syria, where it precipitated civil war. In 2000, Israel came up with a response to water stress, launching a Desalination Master Plan to provide drinking water to its population, which set the target of producing 1,100 million m^3 of desalinated water per year by 2030 and announced five large desalination plants. These plants now supply almost all of the country's tap water. Israel later became a net exporter[134] of water and producer of the cheapest desalinated seawater on Earth.

But nowhere is safe from water scarcity. The American Southwest is experiencing a new era of extreme heat, drought and aridification: drought impacting the Mississippi River coupled with a long history of dredging by the Army Corps of Engineers has contributed to problems with drinking water in populous cities like New Orleans.[135]

Initially, the US invested heavily in desalination technology but, due to rising costs, many plans to build plants were abandoned.

Micro desalination plants are common in the US near natural gas and fracking facilities, and the majority of desalination plants are in California, Florida and Texas. The largest plant of this kind in the US is the Carlsbad Desalination Plant in California. It creates over fifty million gallons of freshwater a day, used by the approximately 3.1 million people in the region, but it cost $1 billion to build and maintain.

Within Europe, desalination plants are primarily located in Mediterranean countries, with Spain being one of the world's largest users of this technology. In Asia, countries such as Singapore, India and China have developed desalination programmes. Australia also has several desalination plants in operation to address water scarcity. A desalination plant refers to a single site or operation, while a desalination programme encompasses the overall effort or system behind deploying desalination.

Case studies of desalination technology are remarkable. The island of Aruba has an average rainfall of fewer than twenty inches per year, so water availability has been taken seriously by this typically dry country, and water needs have been met by desalination since the 1930s when the first plant of its kind was built to convert seawater to drinking water. Desalination now provides the island's only source of drinking water, an impressive feat for a country populated by 100,000 people that attracts over 700,000 tourists every year. Aruba boasts the best water in the world.

Since we have the solution and we know that water scarcity is a growing problem, why aren't countries building more desalination plants? It may not come as a surprise that the main barriers to building desalination plants are cost and environmental concerns.

The waste brine created by desalination plants is a highly concentrated saltwater by-product that contains chemical residues like coagulants. As ever, environmentalists have used this as an objection against proposed projects. For example, in 2022, a twenty-year plan to build a desalination plant at Huntington Beach in California was declined by the California Coastal Commission over concerns relating to waste disposal and the environmental impact of brine.

Since valuable minerals and elements can be extracted from the brine, including gold, silver, lithium, copper, magnesium, potassium and bromine, it could be a valuable resource. Ultimately, a more definitive solution is needed for the waste, which – dare I say it! – is only waste if we waste it.

The other problem is the cost, which is due to the fact that desalination requires plentiful energy. Although energy requirements to desalinate one unit of seawater have been declining during the past two decades, plants are still highly energy intensive. This could easily be addressed by powering desalination with clean and reliable nuclear power plants. One study[136] notes that 'there has been virtually no discussion or appreciation of making desalination infrastructure to be carbon neutral'. While dialogue around the need to power AI data centres moves forward quickly, the discussion around desalination has stagnated for years.

Finally, plants are substantial infrastructure projects, and Western countries are currently not very good at building large-scale facilities that require upfront investment, experienced workers and long-term commitment to building and maintenance. Perhaps that is why desalination is rarely discussed, but since we know that water scarcity will worsen globally, and that water stress drives migration – a top concern in many Western countries – it makes sense to finance and build these plants now in large numbers, especially in areas already prone to water stress.

As President John F Kennedy once said[137] at the opening of the desalination plant in the US, it is 'a work that in many ways is more important than any other scientific enterprise in which this country is now engaged'. It's notable that he said this during the space race. I'm all for reaching for the stars, but to achieve such goals we first need to ensure the very basics of survival for humankind, which means ensuring long-term access to drinking water, especially in regions where the technology is needed most. Let AI data centres have their own nuclear power plants, and let's create enthusiasm for building desalination plants alongside them.

2. We need energy: for air conditioning

As well as clean water, people need access to cool air on hot days. Almost as many people globally die from the heat as they do from the cold; but most of us use central heating systems without guilt, while considering air conditioning in summer to be a step too far. So long as air-con is powered by clean energy, there should be no concern over using them on hot days. It undoubtedly makes our lives better, and in the world's hottest regions it is essential for human health and wellbeing.

Singapore's first prime minister, Lee Kuan Yew, once said:[138]

> 'Air conditioning was a most important invention for us, perhaps one of the signal inventions of history. It changed the nature of civilisation by making development possible in the tropics. Without air conditioning you can work only in the cool early morning hours or at dusk. The first thing I did upon becoming prime minister was to install air conditioners in buildings where the civil service worked. This was key to public efficiency.'

According to the International Energy Agency, of the nearly three billion people living in the hottest parts of the world, only

8% have access to air conditioning. Powering more will save lives and won't undermine a high-energy, nuclear-powered future.

3. We need energy for: data centres

The latest technology panic is over AI, with many theories, old and emerging, over how AI will doom us all. One concern is that AI tools will replace workers and cause mass unemployment, which is likely overblown. Although some jobs will be lost to AI, if history is any guide, new jobs will be created. AI's potential to replace skilled labour is also among its greatest benefits because it can handle complex, repetitive or data-intensive tasks with speed and precision that humans often cannot match. This can free up professionals to focus on more creative, strategic or interpersonal aspects of their work, increasing overall enjoyment and innovation. For example, in fields like medicine, law, engineering and finance, AI can assist with diagnostics, document analysis, design optimisation and risk assessment, respectively, helping experts make better-informed decisions faster. Additionally, by automating routine yet skilled tasks, AI can reduce human error, lower operational costs, and make specialised services more accessible.

More importantly, think of all the regions of the world where children lack access to education, where schoolteachers are scarce and opportunities for adult learning are scant. Think of the preventable diseases that are untreated due to a lack of information, the dearth of health care providers, and how many lives could be improved and saved by overcoming these challenges.

In many ways, AI will be an equaliser for poorer countries where education and health care have historically faced multiple challenges. In fact, a positive AI revolution is already unfolding in the global East and South, in regions that have long faced significant challenges in providing quality education and healthcare due to limited resources, infrastructure gaps and shortages of trained professionals. AI technologies are now offering innovative solutions that help overcome these barriers, sparking a positive revolution that is improving lives and accelerating development.

Education is one major area where AI is making a difference, particularly in countries where millions of children and adults struggle to access qualified teachers or adequate schooling facilities. AI-powered learning platforms provide personalised education by adapting lessons to each student's pace and understanding, a tailored approach that helps students learn more effectively, even in remote or underserved areas, and addresses the shortage of educators.

Healthcare too is being transformed by AI innovations. In some regions, AI tools analyse health data to predict disease outbreaks and improve the supply chain management of essential medicines, ensuring critical drugs reach communities in need more reliably. Telemedicine, powered by AI, enables remote consultations and diagnostics, allowing patients in rural or isolated locations to access medical advice without travelling long distances. This expands access to quality care and supports overburdened health systems.

Agriculture, which sustains many developing economies, is another area where AI is making a significant impact. Sensors and data analytics, powered by AI, help small-scale farmers optimise crop and livestock production by monitoring conditions and recommending precise interventions. These improvements reduce waste, increase yields and improve farmers' incomes, contributing to food security and economic stability.

In Japan, AI helps with the ageing population, where the shortage of care workers is remedied by using robots to patrol care homes to monitor patients and alert care workers if something is wrong. These bots use AI to detect abnormalities, assist in infection countermeasures by disinfecting commonly touched places, provide conversation and carry people from wheelchairs to beds and bathing areas, all of which means less physical exertion and fewer injuries for staff members, plus a higher quality of care. Biometric data gathered from wearable devices might be a game-changer by detecting early signs of cancer, monitoring the spread of infectious diseases and other general health issues, and providing patients with agency over their health where access to care is limited or expensive.

AI already plays a role in helping humanity tackle natural disasters, for example, by predicting an earthquake's strength and how many aftershocks[139] will strike. These models have been trained on large data sets of seismic events and found to estimate the aftershocks more closely than conventional (non-AI) models.

Forecasting models help to predict other natural disasters too, in the form of severe storms, floods, hurricanes or wildfires.[140] Machine learning uses algorithms to reduce the time required for forecasts and to increase model accuracy, which is again superior to non-AI models. These improvements could have a massive impact on the rural populations of poor countries who are frequently employed in agriculture, which is highly dependent on the weather but currently lacks reliable forecasting.[141]

Much of the fear over AI in the West concerns the rapid speed at which it is being implemented but, for many countries, this speed is a boon. Take the mobile phone as a similar example. In 2000, only 4% of people in developing countries had access to them[142] but, by 2015, 94% of the population had phones, including in sub-Saharan Africa. The benefits were enormous, as billions accessed online banking, educational opportunities and reliable communication. One study[143] found that almost one in ten Kenyan families living in extreme poverty lifted their incomes above the poverty line by using the banking app M-Pesa.[144] In rural Peru, household consumption rose by 11%, given access to phones, while extreme poverty fell by 5.4%. Some 24% of people in developing countries now use the mobile internet for educational purposes, compared with only 12% in the richest countries. In lower-income countries, access to mobile phones and apps is life-changing.

AI, which needs only a mobile phone to use, is foreseen to spread even faster in countries that need the technology most. This is where the conversation should focus: not on technology panic, but on a technology revolution for greater equality in wellbeing. And on how to power it.

What's in a name?

In 2023, when Germany's last nuclear reactors were shut down, I spoke at a demonstration in Berlin. On one side of the Brandenburg Gate, Greenpeace was celebrating the nuclear phase-out with a party. On the other side, a group of pro-nuclear activists was lamenting the same event.

I have been described as 'pro-nuclear' for years now, although I previously considered myself to be 'anti-nuclear'. I've been on both sides of the gate, but have come to believe that these terms are meaningless. I think the notion of being for or against any specific technology is a game of false semantics. It all depends on the story, and the storyteller, and I think we've all been duped.

How can we claim to be 'pro-sun' or 'pro-gravity'? These things exist and they influence our lives, but whether or not we like them is irrelevant. Since we don't control the sun or gravity, let's apply this idea to examining our attitude to human-made technology instead.

And what exactly is technology? According to the Cambridge Dictionary,[145] it is 'the study, knowledge, method and application of science'. This broad definition implies that technology covers everything from mattresses to pencils to shoes. It is correct, even if when many people initially think of 'technology' they picture an electrical device.

To make sense of why this is the case we have to revisit our understanding of nature and enter the murky territory of defining what is considered 'natural' and what 'nature' really is. In its simplest form, 'nature' is the world around us. This concept has become muddied in popular use, as the traditional environmental movement has shaped the concept of 'nature' as a special, even revered, entity personified by the old Greek myth of Gaia. According to this narrative, nature is everything that exists around us but was not created by us. In this story, 'nature' is independent of human impact; a world before we existed. But in the reality of scientific understanding, humans arose from the Earth's natural processes and evolution and now create things using materials mined from the Earth; this makes humans a *part* of nature, not *apart* from it.

This leads on to exploring why some people are 'anti-GMO' but not 'anti-organic'; 'anti-nuclear' but not 'anti-sun'. Conflictingly, when people promote solar panels they are endorsing the sun, which through nuclear fusion constantly uses up the hydrogen at its core. Sounds dangerous? I'm surprised people aren't protesting it.

Solar panels are a form of technology. They depend on mining, raw materials and engineering skills to construct. Why are they seen as natural and therefore good, when nuclear power plants that require considerably fewer materials are seen as bad? Because these perspectives have been normalised by a small faction of society. It has become socially acceptable to speak against certain technologies but not others, based entirely on the spurious notion that some technologies are 'natural' (an argument regularly made about wind and solar power), while others are not.

Actually, nuclear fission is a natural process,[146] one that has been found in nature. In the Franceville Basin of southeastern Gabon, scientists discovered ancient natural nuclear reactors, known as the Oklo reactors, which operated approximately two billion years ago. These reactors formed when uranium deposits underwent self-sustaining fission reactions, moderated by groundwater, without any human intervention. The conditions for these natural reactors included high concentrations of uranium-235, a neutron moderator (water), and a lack of neutron-absorbing materials. The reactors operated intermittently over hundreds of thousands of years, producing modest power outputs of about 100 kilowatts, sufficient to power approximately 1,000 light bulbs.

Arguably, that makes fission more natural than a kettle or a toaster. What about pencils and mattresses? We don't consider these to be technologies because they are not contentious 'wedge' issues with people protesting them. We don't have to be 'pro-pencils' because there is no loud and powerful activist movement against pencils. We simply accept that pencils are probably useful and better our lives somehow.

The terms used to describe certain technologies have been defined by activists and hippies, mostly out of negativity bias. But that is not what science shows us. When asked, most people will not say that they are 'pro-shoes' or 'anti-nuclear'. Instead, they support broader ideas: no, of course we cannot walk outdoors barefoot; yes, of course I need to use lighting and to charge my phone (and I'd like to be able to do that reliably and inexpensively). Ergo, they inadvertently, and naturally support things like shoes and lighting. And if better and improved technologies come along, they will support them too.

Considered in this way, I am not really 'pro-nuclear'. There is no such thing. The technology exists. The technology has immense potential. My stance is irrelevant in relation to whether or not it is necessary. Reliable energy is necessary, to escape poverty, warm our homes, light our streets. Clean energy is necessary to improve and save lives. Being for or against it does not change these basic truths.

Arguably, if something is invented and improves your life in some way, then you are already 'pro' it, since you use it. Broader society, perhaps even universally, has approved it. For example, everyone is pro-soap. Pro-cooking. Pro-flush toilets. I have worn glasses since I was nine years old due to myopia. Do I consider myself to be pro-glasses? Do you? Does anyone? (Actually, watch that space, because there was a recent video of a woman claiming that we don't need to wear glasses.[147] Everyone on the internet appeared to laugh at her. But how is being anti-glasses different to being anti-nuclear or anti-GMO?)

The idea of being against something that is not only part of quotidian life but possibly essential to it, came from activists. Within all the pushback against doomers, we have fallen for depressing rhetoric. The narrow thinking of a limited few has had a significant impact on the way we all think and talk about technology.

To be fair to activists, there have been positive changes in the world thanks to campaigners on single issues. When I refer to 'activists' here I am not referring to small groups of well-meaning campaigners, but to large, well-funded and influential interests with questionable motives, for example NGOs that have worked to undermine nuclear energy in favour of fossil fuels,[148] and those that have damaged energy security by working with dictators.[149] These groups have made some of the challenges we face today, like building clean energy, much harder than they should be. The damage they have done is incalculable.

Historically, debates involving technology have lacked balance. Journalists initially didn't balance coverage of anti-nuclear protests with the voices of people in favour of the technology, because originally there were no 'pro-nuclear' people. There were only engineers, physicists and other scientists getting on with their jobs, who did not identify with the term 'pro-nuclear'. It would be like calling themselves 'pro-physics'. So, they had no voice in the debate and their views, based on facts and data, were not represented. Only the *feelings* of activists were heard. For decades, there was no debate at all. Eventually, when 'pro' voices entered the chat, the discussion began to move forward.

Now, to have a voice in the debate, people increasingly choose to identify as 'pro-nuclear', but this is still part of a wider deceptive narrative. The narrative has been repeated for so long, across so many mediums, that it has spread a sneaky feeling that technology is bad: that it comes for our jobs in the night, disconnects us from each other and from nature, and it has ruined, or will ruin, the world.

Consider past technology panics about calculators, microwaves, mobile phones and even coffee. They sound absurd to us now. Yet others remain: panics over nuclear energy, GMOs, AI and so on. These are not political issues. They are discoveries and inventions that have been made political. Being against them is as absurd as protesting coffee.

Technology is only a tool. How we use it is up to us. We can be in favour of using it in a specific way, but that's just called having an opinion. Being in favour of the fact that it exists, or not wanting it to exist? That's bizarre. And yet those are the terms we have been reduced to: 'pro-nuclearism' and 'techno-optimism'. As if logical alternatives to these positions could possibly exist.

Everything around us is technology. I wear shoes, glasses and clothes. I use a mobile phone and laptop. When I am in 'nature', am I free of technology? Let's remove the shoes, glasses and clothes. Now we are back to the old hippy idea of being 'one' with nature – naked Woodstockers rolling around in the mud. It was the Sixties, man. But we have to move on from that reductive way of thinking and from that era. If I pick up a stick on the moors and use it to start a fire, didn't I just create a tool out of nature?

Marc Andreessen's *Techno-Optimist Manifesto*[150] is an impressive and lengthy missive that sets out his perspective on the role of technology (and also incorporates a phrase I helped to popularise: 'energy is life'). I have some admiration for the ambitions and future-gazing that Andreessen's manifesto represents, but it also perpetuates the narrow concept of what constitutes technology and frames it as something that we have to choose to be for or against. The premise is based on anti-technology arguments.

There is no such dichotomy. It's just semantics. Technology itself is value-neutral. Do I want to see positive applications of new technologies? Of course. But that's like saying that I'm pro-breathing. Aside from a tiny number of sociopaths, don't we all want to see positive applications of technology? Even the most Luddite of Luddites. What are all those dystopian post-apocalypse films and books about, if not the dread of negative applications of technology and the desperate wish to avoid them occurring?

In broad terms, this makes us all techno-optimists (barring the few sociopaths). Back to the false narrative I mentioned earlier. The lie goes deeper than being for or against technology. The lie is that there is a binary choice between being pro- or anti- any form of technology that we already depend upon. Technology is a part of us; we are already cyborgs in that regard. Without my glasses, I wouldn't be able to write this article. Without my shoes,

I wouldn't go outside. Without coffee, you'd probably be asleep instead of reading this.

As long as humans exist, so will technology. To say that I am 'pro-nuclear' is no different than saying that I am 'pro-Earth' or 'pro-oxygen'. All these things are true, but saying them implies that there is a valid contrary stance, which is incorrect. A person can hold any opinion, but it is only valid if it is logical or based on fact or scientific truth. People can be 'anti-breathing', but let's see the longevity of a movement formed on that basis.

We all live with the basic fundamental laws of the universe. In this world, fission exists, whether people are for or against it. Whether or not we use it well, to improve lives and better humankind, is up to us. A few people who might protest shoes as vehemently as they protest nuclear power plants have held centre stage in this debate for far too long. Journalists need to stop amplifying their voices. And we need to start adopting better terminology ourselves. While labels can be powerful, they can also be limiting, and certain terms that we have become accustomed to when discussing technology are redundant.

The language we use matters; it shapes the way we think and helps to configure society. If we continue to define our ideas based on oversimplified framing developed by Luddites, we will continue to limit ourselves. By defining ourselves in opposition to them, we allow their ideologies to hold us back. The real challenge is to move away from traditional framing and, like the idea of thinking outside the box, learn to approach things from outside the Gate.

The question of how much the language you speak affects the way you think has been debated for a long time. The linguistic relativity hypothesis suggests that the grammar and vocabulary of a language influence the way its speakers perceive and interpret the world. Then there's Shakespeare's philosophy that 'a rose by any other name would smell as sweet', which suggests that language does not determine thought but when ideas are represented clearly by words it enables speakers to consider those concepts more easily.

How does this apply to talking about the environment? Language can make us think differently about things, and activists are great at using it to achieve their aims.

Reframing arguments involves changing the way an issue is presented or described to shift people's perceptions and responses. By using different language or emphasising alternative aspects, reframing can make a topic more relatable, less threatening, or more appealing to a specific audience. This technique taps into the psychological power of framing effects: how people interpret information based on context, which can be especially effective in debates, negotiations and persuasion, by helping listeners see the issue from a new perspective, breaking down resistance and encouraging open-mindedness.

Within a few months of the XR messaging team promoting our new terminology, governments around the world declared 'climate emergencies' and set 'net zero targets'; newspapers announced that they would use less passive language when reporting on climate change and adopt the terms 'crisis' and 'emergency' instead of 'climate change'. Even *Scientific American* published an article explaining why it decided to adopt the term 'climate emergency'.[151]

In some ways this was a much-needed shift: it gave people agency to talk about climate change in a way they couldn't before, when they had only scientific jargon at their disposal. Facts don't always speak for themselves, so boiling down ideas to simple concepts can achieve much more than making someone read scientific papers. Disinformation campaigns have played a part in misleading people for decades – from frightening them against nuclear energy to confusing them about climate science

– therefore, it's unsurprising that a new story being written about climate change was well-received by journalists and politicians.

However, what hasn't been debated enough is the purpose of this change of language. Famously, in 2019, activist Greta Thunberg gave a speech at the World Economic Forum in Davos, stating: 'I don't want you to be hopeful. I want you to panic. I want you to feel the fear I feel every day.'

Activists who popularised those terms intended to alert and alarm people, even to worry and frighten them. They pushed for degrowth and living with less. How could that solve any of the world's problems? People were quick to adopt the language, but few asked whether it was the best language to adopt. And others, like Andrew Neil, pushed back on claims that billions would die, which only weakened our overall arguments.

It should be of concern that the language we now use to describe important global issues comes from a small group of activists who believe that humankind is lost, who often fear technological solutions and suffer from technophobia, and who also believe in degrowth and long-debunked theories of overpopulation. It hailed from the same activists who tell stories of three-eyed fish to shut down nuclear power plants, misinform people against life-saving genetically modified organism technologies, and scare parents against vaccinating their children.

For many decades, non-governmental organisations and social movements have made strategic use of science to promote their ideological stances and influence political and economic decision-making. Unfortunately, much of the language and storytelling used has been dystopian, frightening and against scientific consensus. The crux is that the ideology behind words that alert and alarm us is a belief that humans are bad, that we got it wrong, that we are to blame for environmental destruction, for which we will be punished. This is a tale of original sin from the traditional environmentalist, whose god is whatever is perceived as 'natural', a term that is entirely meaningless since nothing is truly natural (or everything is). Natural disasters are often deadly to life on Earth. Childhood diseases are technically natural. Weather is natural; so some people consider solar and wind energy good but nuclear energy bad (despite the fact that atoms make up everything and nuclear fission has been found to exist independent of human influence, in nature).

These ideas have influenced people widely, even those who do not consider themselves environmentalists. Many people feel responsible for what they believe to be the negative state of the world and feel guilty about their lifestyles; about using electricity, or even about having children. They feel anxious about the future.

In the study of Science Communication, activists are known to draw on the 'symbolic legitimacy' of science to attain credibility, while making strategic use of science to achieve their aims.[152] As the authors note:

> 'Activism can also oppose or even threaten scientific legitimacy. Civic protests against potentially risky or ethically contested scientific or technological developments such as genetic engineering, nuclear research or nanotechnology are relevant examples. The strategic (mis)use of science or the use of counter science by social movements (e.g., anti-vaccine campaigns) draws attention to the critical role that activism can play for science communication.'

This is a dangerous game, because the environmental movement that idolises poverty and abhors progress has led the charge with rebranding environmental messaging. Rebranding and storytelling are closely linked because both involve reshaping how people perceive a person, product or idea, through carefully crafted narratives. Storytelling provides the emotional context and meaning behind a brand, turning it from just a name or logo into a relatable, memorable identity. In essence, rebranding is storytelling in action: it uses narrative techniques to rewrite the brand's story, influence perceptions, and build trust and loyalty. Without a compelling story, rebranding risks, feeling hollow or confusing, but with strong storytelling, it becomes a powerful tool to transform and engage audiences.

Humans experience many cognitive fallacies and emotive language that is upsetting, and it is more likely to impact us and stay with us. We enjoy drama; it taps into our emotions and social brains, making stories more engaging and memorable. When we watch or hear dramatic situations, our brains release dopamine, the 'feel-good' chemical that keeps us hooked. Drama also mirrors real-life conflicts and relationships, helping us explore complex feelings and social dynamics safely from a distance. It can trigger empathy, suspense, excitement or even catharsis, which is why drama is such a big part of storytelling across cultures and throughout history. In short, drama satisfies our innate curiosity about people, conflict and resolution.

We believed the stories, and we absorbed the message when we were told to panic about climate change. The terms 'crisis' and 'emergency' became widely popular, but have they solved anything? Were they intended to?

The language we now use to talk about climate change also fuels eco-anxiety in children: a 2021 survey[153] of 10,000 young people between sixteen and twenty-five years old, across ten countries, found that more than 50% of respondents reported feeling sad, anxious, angry, powerless, helpless and guilty, and 56% felt that 'humanity is doomed'.

Language may not shape our worlds as much as some linguists

previously believed, but it does affect the way we respond to current challenges.

Doomerism is not going to solve climate change, end poverty or tackle air pollution. It will not lead to building more clean energy or air conditioning units to cool people through hot summers. Language that alarms but does not inspire may not be as helpful as many media outlets and world leaders have been led to believe.

In promoting the benefits of nuclear energy, I have tried to tell positive stories about clean energy and inspire people to imagine living with an abundance of energy[154] rather than with scarcity, popularising terms like 'Nuclear saves lives'; 'Nuclear workers are climate heroes'; 'The most dangerous reactor is the one you don't build'; 'Nuclear energy is clean energy'; 'Nuclear is zero carbon' and so on, to reframe nuclear energy. Fighting against the tide has been successful in many ways but has not yet been enough to propel the world into rapid action.

For example, roughly 90% of homes in the US are air-conditioned,[155] compared to only 5% of European homes.[156] With hot summers across Europe, and now increasing heatwaves, mere talk of crises or emergencies will not make air conditioning units more affordable for homes and businesses. Nor will advocating for degrowth. Yet indoor temperatures, even in climates that are heating up, are things we can control easily. There are life-saving actions that we can carry out now, but few politicians, activists and non-governmental organisations mention this simple fact, despite using strong language about the need to act. And to keep all those air conditioners working will take plenty of energy, so we should plan to build enough clean energy to meet demand.

When we think of the future we want to build and the world we want our children to inherit, we need to use language that points to solutions and to ensure that the ideology behind the language is not technology-averse or apocalyptic. With positive language, we move away from dystopian visions of the future and instead tell better stories; ones about our past as well as the future. Yes, we made some mistakes, but we also escaped poverty, and we can continue to make the world a better place.

Say we have been told that our house is on fire. Since we are in this emergency, we need logical thinking and practical steps to combat the problem at hand. Anxious about a heating planet? Nuclear power plants are your metaphorical fire engines. Build a lot of them. Worried about excess heat? Manufacture air conditioning units. Power them with clean energy. Troubled by rising sea levels? Build sea defences and flood barriers. Thankfully, all the solutions are already available to us. There isn't a challenge we can't rise to if we want to protect as many lives as possible from harm.

We can build a high-energy future and live energy-rich lives without causing planetary destruction. We can end energy poverty, eradicate air pollution, and cool the air in our homes, schools and places of work so that heat-related deaths become uncommon. This may be an ambitious vision, but so what? Dystopian horror stories have consumed enough of our waking hours. To make climate-adapted prosperity a reality, we need to use the language of solutions and re-evaluate the merits of heating up language that alerts and alarms people. Perhaps, instead, like the world around us, our language needs to chill out.

CHAPTER THIRTEEN

WHAT ARE THE CHALLENGES OF A NUCLEAR-POWERED FUTURE?

'Any technological advance can be dangerous. Fire was dangerous from the start, and so (even more so) was speech – and both are still dangerous to this day – but human beings would not be human without them.'

Isaac Asimov

Fearful narratives aside, the two most commonly raised issues regarding growing nuclear energy are construction time and financial cost. These topics are intertwined, and they are also political rather than a technological issue. Let's address why that is.

Countries like Japan, South Korea and China have built nuclear reactors in around five years. Newly-built South Korean nuclear plants have also achieved a speed and cost-effectiveness in construction that exceeds most other countries. According to an energy policy paper,[157] the decline in South Korean nuclear power costs compares to the decline in solar power costs in Germany over the same period. Building nuclear power plants in China is seven or eight times cheaper than anywhere else, where twenty-one nuclear reactors are currently under construction (two and a half times more than any other country) and each reactor costs roughly $2 billion to build roughly over those five years.[158]

The slow speed and correlative high cost of nuclear energy can be attributed to three dominant factors: over-regulation, failure to standardise and loss of skills. So, how can we do it differently? What do we need to change? Let's begin with the pitfalls of over-regulation.

Some regulation is necessary with any technology, but nuclear construction regulations across the West have become cumbersome due to decades of over-focus on safety and environmental standards, largely thanks to environmental groups who protested new nuclear construction to deliberately drive up costs and build times by lengthy regulation processes. Environmental standards are positive when they're necessary, but sometimes they're misguided and hold up progress.

For example, an anti-nuclear activist group in Britain that campaigns against the proposed new reactor at Sizewell C nuclear power station, launched and lost a legal challenge against the energy provider EDF Energy. The anti-nuclear group argued that putting a new reactor at the existing power plant site is ecologically unsound. Yet the environmental statement that the Sizewell C team put together in 2020[159] is 203 pages long and demonstrates a net gain for biodiversity from the site. EDF's environmental plan includes a new eleven-acre marsh harrier wetland habitat for wildlife,[160] a mosaic of wet reedbed with 20–30% open water and 1 km of lowland ditches.

A year after they launched the legal challenge, a judge dismissed it, stating that the group's arguments were 'totally without merit'.[161] For the anti-nuclear group, the year of stalling, the required paperwork, a two-day hearing and an ongoing appeal process, amounted to a victory. This process alone had already added to the cost and build time of Sizewell C.

Because of stunts like this, despite cross-party and public support[162] for nuclear energy in the UK, it's hard to get new infrastructure projects off the ground. People lose hope in the technology when they see constant delays without knowing the cause. This has led to the widely believed myth that nuclear power plants are inherently slow to build. Last year, the think tank, Britain Remade, highlighted a case where a railway restoration project near Bristol generated 79,187 pages of paperwork that, if laid end to end, would be 4.5 times longer than the proposed railway line.[163]

So you see, I am not arguing against regulation altogether, but for us to look at where it has become a barrier to progress for no real reason. This is not just a nuclear-related problem but appears to affect many areas, from 5G technology to childcare. Research has found that over-regulation has potentially prevented[164] the US from being world leader in 5G technology, causing it to lag behind China, and it has even made childcare more expensive without improving it.[165] In the UK, cumbersome and slow planning process delays much-needed new housing; regulation prevents crucial clinical research and erodes trust in doctors.[166]

Meanwhile, South Korea's APR-1400 reactors are highly reliable,[167] with three times fewer capacity losses than in the US and six times lower than the UK's reactor fleet; but if the Korea Electric Power Corporation (KEPCO) wanted to build a nuclear power station in Britain it would take considerably longer than the few years it takes at home.

Firstly, the company would have to complete an assessment, submitting the design to the Office for Nuclear Regulation (ONR) and the Environment Agency (EA), to be assessed for safety, environmental impact and waste management; this stage takes around four years. Next, they must apply for planning permission, after which the application needs to be approved by the local council, which means gathering the support of residents and regional politicians. Then, even if KEPCO negotiated the cumbersome regulatory process, there's no guarantee that the reactor would pass or that the company wouldn't have to make costly design changes, which is a disincentive.

Although certain regulations have made nuclear energy safer, most act as bottlenecks that prevent us using this life-saving technology; the lengthy time constraints and mounds of paperwork required by the regulatory process turn off investors in droves.

In 2019, the Japanese company, GE Hitachi Nuclear Energy, lost £2 billion after failing to secure the necessary government commitment to build at the Wylfa site in Wales and ultimately gave up on the project. The greatest irony is that standards for nuclear builds are stricter than those for building coal-fired power stations, and claimed to be in the name of safety. We have anti-nuclear activists to thank for that, as planning regulators claimed that it would 'adversely affect Welsh language and culture'. Yet, when I spoke to the community in Wales, they seemed one of the most pro-nuclear populations in the world, and expressed worries about young people leaving Wales due to the lack of jobs since all local industry has shut down, leaving only the current reactor that is being decommissioned. Working-class voices are seldom heard.

New reactors under construction in the UK are European pressurised water reactors (EPRs). However, French energy company EDF's reactors only gained approval in the UK after making thousands of changes to the reactor design used in France. People like to blame the EPR reactor design, but as *The New York Times* reported:[168] 'EDF placed additional blame for the delays and cost overruns on Britain's nuclear regulations. To meet the requirements… the original design would need 7,000 changes, including 35% more steel and 25% more concrete'.

As well as being required to source and pay for extra materials, without which the reactors function perfectly well in France where almost 80% of the country's electricity comes from nuclear energy, this also means that lessons EDF have learned about reactor design cannot easily be applied in the UK. The changes disrupt the process of standardisation, causing further delays as the reactor is essentially built from scratch in an amended way from the original fleet in France. That is a problem because productivity gains come from practice and experience. A simple question is why do all nuclear reactors built in the UK need approval from the ONR, when South Korea's APR-1400 has already been approved as safe in the US and parts of the EU?

Medicines recognised as safe in the EU, US or Japan are often automatically approved as safe in the UK. We do not test them according to our standards; we trust this has already happened in reliable overseas locations. Why not apply the same approach to nuclear reactors? If we were to automatically approve designs accepted as safe elsewhere by respected regulators like the NRC, this would shorten the time taken to bring a South Korean reactor to market, while avoiding unnecessary design changes, additional costs and build times.

In the US, a new reactor must be approved by the federal Nuclear Regulatory Commission (NRC), a process that takes up to five years, inevitably increasing costs. Twelve states have additional independent conditions[169] restricting new nuclear construction. Some regulatory increases to address safety concerns after the Three Mile Island incident were reasonable, but many had nothing to do with safety. Several countries that have safely operated multiple nuclear sites for years show how regulation can engender safety without holding back vital technology. We should entrust building large infrastructure projects to engineers, not lawyers. Western countries would gain maximum benefit by re-regulating nuclear power to build nuclear power plants quickly and at cost.

One of the reasons South Korea's costs are low compared to the UK is because they build whole fleets of reactors rather than one at a time. Every time they build a reactor, they learn something new and can apply their experience to future projects.

The same team builds the same reactor repeatedly, which means they get better and faster. This is efficient standardisation or 'learning by doing'. This standardisation has lowered the building costs[170] of nuclear power plants in France and the US; meanwhile, the UAE's Barakah plant employed the same approach and succeeded in building four reactors in under a decade. Between 1970 and 2000, France built fifty-eight nuclear reactors at an average of 2.6 years faster than US plants, due partly to a high degree of standardisation.

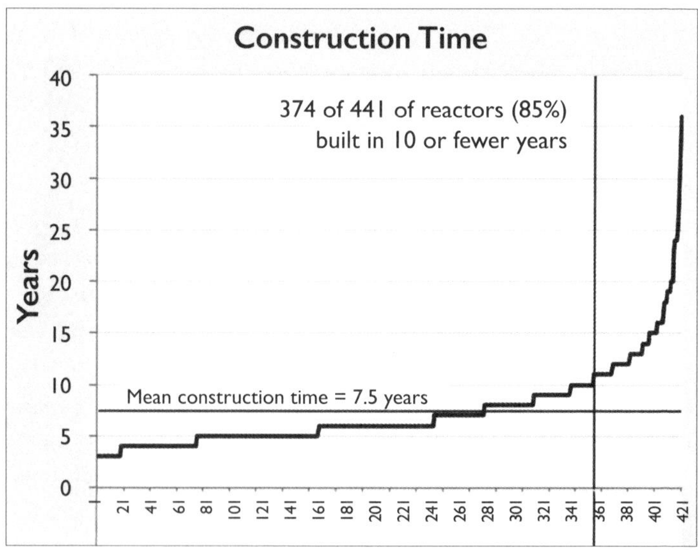

The economic argument for nuclear energy, despite the price tag, is well supported by data. An IAEA report shows that investments in nuclear power produce the most considerable economic multiplier effect of any clean energy source.[171] Nuclear power creates about 25% more employment per unit of electricity than wind power, while workers in the nuclear industry earn one-third more than those in the renewables sector. These jobs are well-paid, long-term and predominantly local.

A report[172] by the British Science Association, assessing insights into young people's attitudes towards nuclear energy, found that young people in Britain are open to careers in nuclear energy but don't feel informed about it. The research showed that fourteen- to eighteen-year-olds felt more educated about renewables and fossil fuels than nuclear energy, including through schooling. It showed young people to be highly concerned about climate change, and wanting to learn more about how nuclear power could help to address it.

Almost two-thirds (64%) of young people said they would be interested in careers in nuclear energy, and 70% believed that a career in the nuclear industry would be 'challenging' rather than 'uninspiring' (6%). 58% considered such a role would be 'interesting' rather than 'boring' (15%). These are substantial statements that, if leveraged, could create the nuclear workforce Britain needs to get back on track energy-wise.

Individuals who care about the planet's future and want to make a difference will achieve more by working in the nuclear sector rather than blocking roads. Ideally, our formal education process should inform them of this, but it is currently woefully inadequate when teaching electricity generation and climate solutions. In Britain, adding these subjects to the curriculum would at least address the problem.

In France, where the nuclear energy is the third-largest industry after the aeronautics and automotive sectors, the public already has a positive view[173] of vocations in nuclear energy production. However, as with other Western countries, France suffers from a skills mismatch, and France's manufacturing – construction, engineering and IT industries – struggle to obtain the workers they need. The French government is attempting to entice and recruit workers to build power plants and has launched a construction drive unseen in nuclear science since the 1970s. This includes a disturbance allowance equal to two months of salary for relocation, a 'discovery pack' to help workers familiarise themselves with the location they will be working in, a financial 'mobility pack' to assist in finding accommodation, employment for spouses, and benefits for childcare and schooling.

Globally, many countries face the challenge of recruiting more people into STEM vocations. One simple step to help remedy this is to remove training and qualification barriers by offering apprenticeships, removing loans and fees, and guaranteeing work after the study period by committing to new infrastructure projects so that jobs are readily available. With nuclear, there is the additional issue of people not knowing enough about it to want to train in a related field in the first place.

Unless you're building a small modular reactor, there's no need to change existing large-scale nuclear reactors. Materially-efficient reactor designs already exist, and can be built quickly. There is, however, room for innovative ideas that might shape the future of new nuclear projects, such as 3D printing parts[174] for nuclear plants, transitioning coal sites to nuclear sites,[175] and improving efficiency at power plants by employing AI tools.[176]

The required changes may seem complicated, but in practice, they are straightforward. Pick a good design, remove regulatory barriers, check the supply chain, inform and incentivise workers, and proceed to build fleets at a time. It's not rocket science.

We can reset this burdensome approach to building nuclear power plants. Imagine a hyper-efficient way of building reactors quickly: a gigafactory, printing nuclear reactors one after the other. It remains to be seen whether any country will propel humankind forward by setting and achieving such ambitions, but a future of energy abundance and resilience is up for grabs if we choose to permit it.

A remaining issue is not one of nuclear ability but the likelihood of whether or not we choose to wean off of fossil fuels completely, as so many now demand. The irony is that those who most want to wean off of fossil fuels have had the most comfortable lives of all. They haven't suffered through truly gruelling labour, have no need to understand how electricity grids work, and have rarely stepped foot in a mine or experienced electricity blackouts. The divide between what it takes to maintain a high quality of life and the scientific understanding of it has widened over generations.

I'm a huge fan of the late science fiction writer Isaac Asimov. In his *Foundation* series, Asimov is concerned with the stagnation and degradation of society once a certain level of comfort and wealth has been achieved. One of the main characters, mathematician, Hari Seldon, tries to develop a way to predict future downfalls of humankind and address them before they take place through a method he calls 'psychohistory'; this is a mathematics of sociology that he develops in the books.

The most unusual thing about the *Foundation* novels, and Asimov's works in general, is the timescales he works with. We are not talking about looking a few decades into the future, but tens of thousands of years ahead. Seldon is eventually able to predict the imminent fall of the Galactic Empire, and to predict actions that may prevent a Dark Age from lasting 30,000 years before a Second Empire arises. He puts wheels in motion to avert these events, or, where they seem inevitable, to ensure that there is a push to rebuild afterwards.

Asimov's long-term thinking throws into sharp relief just how short term today's approach is. Politicians don't generally think beyond the end of their term of office, let alone decades into the future. Society commonly does not think of the future past their children or grandchildren. This lack of prescience is short-sighted.

I believe that, in the West, we are currently experiencing a similar dilemma to what Asimov portrays; although our society is less advanced than his, we have become so comfortable with our achievements that we are complacent about maintaining them or progressing further. As evidenced by constant criticism of how long it takes to build nuclear power plants (the average build time is 7.5 years even though they can provide electricity for the entire human lifespan), we are bad at building for the future and do not plan long term for upcoming generations.

Thanks to years of development and growth, we live in a time of prosperity, facing fewer core challenges than past generations. Need to read after dark? Flick a switch and you have light. Want to go for a walk? Great, the streets are lit too. Feeling sick? Call a qualified medical professional. And so on. The problem is that, thanks to this privilege, a disconnection has occurred between our lifestyles and what it has taken to arrive here. Lack of understanding of the skilled labour needed for these industries, the mining for raw materials that enabled our lifestyles, and the fact that the toll those things took on the environment was inevitable (but not permanent), has repercussions for society. I was once part of the problem, fighting to forge a society independent of fossil fuels, and fighting for the alternatives without thinking about our continued and growing need for reliable electricity. My argument was the standard traditional environmentalist argument – that we need to live with less. I was wrong.

Something I have not changed my mind on is coal burning. In 2008 I attended a protest camp organised by Camp for Climate Action. It was our third protest camp, which involved squatting on public land and living there for a week, culminating in a direct action at the end of the week. In many ways it was idyllic: we lived communally, made decisions using consensus-based decision-making, we camped and cooked together, and hosted workshops on everything from gardening to non-violent direct action training. It was family friendly and previous camps had had a positive atmosphere. That was not to be the case at Kingsnorth.

At the time, although the power station at Kingsnorth in Kent was due to be closed down and demolished, there were proposals to build a new coal-fired power station on the same site. This is what we were protesting, but far from being a relaxed hippy camp for the week, we were subject to police brutality with excessive stop and searches, as well as being harassed and intimidated. I recall being woken up in the night by screams of 'Cops on site!' and police dogs barking, as policemen trampled tents and hit people with batons under the cover of darkness, setting dogs to run unceremoniously at us. The tactic was to frighten us into wanting to leave, or keeping us off site by arresting then releasing us without charge, but insisting we could not return to the camp.

The friend I was with was arrested for carrying a knife and fork, which he used to eat his meals. I was later arrested on suspicion of colluding to organise violence. It was an awful, scary experience and later some protesters received compensation for the way they were treated by the police.[177] In this case, there was some truth to the challenges of going up against The Man.

But, whether by our influence or not, the new coal-fired power station did not go ahead and I do not regret my involvement in that protest, though I was somewhat traumatised by the brutal treatment of my friends and being arrested then released in the middle of nowhere without a phone and told to leave the country.

All of this is to say that I stand by our stance on coal then, and I would do it all again if I had to. It is possible to accept two truths even if they feel conflicting: that coal has been a major source of energy for centuries, powering industries, homes and electricity grids around the world, and lifting countless citizens out of poverty. But also, burning coal comes with serious downsides that affect both our health and the planet's future, and it needs to be replaced.

When coal burns, it releases large amounts of carbon dioxide (CO_2), a greenhouse gas that traps heat in the atmosphere. This leads to global heating, causing more extreme weather events like hurricanes, droughts, floods and wildfires. To avoid the worst effects of climate change, scientists agree we need to drastically reduce CO_2 emissions, which means cutting back on coal.

Burning coal also releases harmful pollutants such as sulphur dioxide (SO_2), nitrogen oxides (NO_x), and tiny particles called particulate matter. These pollutants contribute to smog, acid rain and respiratory diseases like asthma, lung cancer and heart problems. Communities near coal plants often suffer from poor air quality and higher health risks as they can't afford to escape the smog; and they are always the poorest communities.

But when it comes to other fossil fuels, the answers are less clear.

Just Stop Oil activists argue to 'stop oil' overnight. But can it be achieved? And is it possible without causing immense harm?

We live in a highly oil-dependent society. About 45% of a typical barrel of crude oil[178] is refined into gasoline, and an additional 29% is refined into diesel fuel. The remaining 26% is used to make plastics and other products. There is good news in these figures, as it means that we can reduce our dependence on oil by switching from petrol and diesel cars to electric vehicles, a trend that is already taking place in many countries.

But it is an incremental change that will take time; the opposite of just stopping oil overnight. The reality is that people would be harmed by taking such drastic measures. As we learned from the energy crisis last winter, energy scarcity is dangerous and claims lives.

The bad news is that oil is very difficult to replace. Many well-meaning companies have been trying to go oil-free, such as the Danish company Lego, which manufactures plastic construction toys. Over many years, Lego has spent millions attempting the switch to making its trademark colourful bricks from recycled plastic bottles from PET instead of plastic. Unfortunately, it turns out that the non-oil-based alternative material demands extra ingredients for durability and greater energy for processing and drying, which makes it more carbon-intensive. Tim Brooks, Lego's Head of Sustainability, summarised:[179]

> 'It's like trying to make a bike out of wood rather than steel… In order to scale production [of recycled PET], the level of disruption to the manufacturing environment was such that we needed to change everything in our factories. After all that, the carbon footprint would have been higher. It was disappointing. Instead of going ahead anyway and virtue signalling that they are now plastic-free, Lego has ditched plans to switch to PET,[180] since doing so would have greater negative impact on the environment.'

Of course, anti-oil activists might argue that Lego should stop making the bricks altogether; that they are unnecessary and wasteful. But I disagree. I recall many years of constructing large and complicated architectural designs with my brother when we were children; he was fascinated by the way things work and are put together, and he later trained to become a senior engineer. This link between playing with Lego and developing engineering skills has been well-documented.[181] The irony of wanting to 'just stop oil' is well represented in the idea of not 'needing' Lego. We don't 'need' to build housing, railways or power plants either – if we are happy for society to stagnate and for future generations to suffer.

Let's ignore the idea of an immediate transition, and consider what can be done to stop oil eventually.

There are many technological solutions to global problems at our disposal today, but change tends to be slow in this area. Politicians are reluctant to invest heavily in large-scale projects or commit to long-term endeavours, while many traditional environmentalists reject what they see as 'techno-fixes', maintaining that it's technology's fault that got us here. I can agree with this statement in a way: without technology, I wouldn't be here, and it's likely that you wouldn't either.

The best option for replacing oil involves pyrolysis. Pyrolysis is the heating of an organic material, such as biomass, in the absence of oxygen. Pyrolysis oil, sometimes also known as bio-crude or bio-oil, is a synthetic fuel that can replace petroleum. It is obtained by heating dried biomass in a reactor without oxygen at a temperature of about 500°C (900°F), with subsequent cooling.

If pyrolysis can replace diesel, why didn't you already know about it? Because it is expensive – partly because of the parts required to construct reactors, and partly because it consumes a lot of electricity. Which returns me to a point I've made for years: we cannot do all of this without abundant clean energy.[179]

As we shift towards electrifying everything, switching from gas boilers to heat pumps, and diesel cars to electric vehicles, much of our demand for oil will be replaced naturally. But the alternatives consume more electricity, and unless the electricity grid is supplying us with clean power, we have a problem. Replacing oil through pyrolysis is part of the solution, but will require more energy.

Social movements often succeed when they choose to fight against, instead of in favour, of something. But I think environmental groups got this approach wrong when it came to tackling climate change. As someone who used to protest fossil fuels, I think we missed a trick in not fighting concurrently for what we need to replace them. Through a combination of misinformation and pop culture references, we even protested one of the key solutions to saving the planet – nuclear energy.

The solutions are in many ways very simple. We have the knowledge and wealth to build what we need when tackling the problems we face. We may lack the skills, due to a shifting workforce, but this can be remedied through offering funded training opportunities, recruitment drives to attract workers and, significantly, the promise of good wages afterwards. Change does not happen overnight, but it can occur rapidly through incremental changes and forward thinking.

When we stop aspiring for progress, we lose sight of who we are. In the middle of the nineteenth century, Britain had less than 2% of the world's population, but was responsible for half of global cotton textile production and over 65% of all coal mining. Now, the birthplace of the Industrial Revolution struggles to build high-speed rail, housing and power plants. To get out of this rut, we could start thinking like Asimov.

When I first diverged from traditional environmentalism, calling out unscientific approaches and advocating instead for solutions, not a day went by where I wasn't accused of being a 'shill'. This is a derogatory term referring to someone who promotes or endorses a product, service, idea or person in a deceptive or dishonest way because they are secretly being paid to do so. It's usually used negatively to suggest someone is acting as a fake supporter or promoter to influence others without disclosing their true motives.

To be clear, I don't work for the nuclear industry. I earned my stripes in the environmental movement and I still believe that environmentalism can usefully contribute to the world, so long as it shifts towards embodying logical reasoning based on data.

Just as psychohistory enables Hari Seldon to predict the decline of civilisation, it can also take steps towards its preservation. We can too. If we truly want to 'just stop oil', we could choose to 'just stop dithering', and 'just start building'. For humankind to flourish, it will take a lot more energy to achieve our aims. We need to 'just start nuclear'.

CHAPTER FOURTEEN

POWER TO THE PEOPLE

'Science and everyday life cannot and should not be separated.'
Rosalind Franklin

We call our planet Mother Earth. It is, as Carl Sagan once put it, the only home we've ever known. It is home to all my friends, family members, co-workers, people I have loved and been loved by, enemies, all the stories I've ever heard, every being I have yet to meet; every thought I have had has been a product and part of this planet. And yet, while our lives unfold here on this small world, our stories and experiences are threads woven into the vast tapestry of the cosmos – shaped by the same stars, the same atoms, the same universal forces that govern the galaxies beyond. Our perspectives are both intimately local and profoundly cosmic, reminding us that though Earth is our only home, we are also citizens of the wider universe.

One day, I will leave this world to my children. I seldom speak of my children in public as I have always wanted to protect privacy, but when I talk of cherishing life and saving lives, I am of course thinking foremost of them.

I do not see science and nature as binary. The fractals found in nature and the chemistry of an apple picked fresh off the tree in my garden go hand in hand. Human distinctions are useful when studying a particular object, but useless beyond that.

The first time I saw the Milky Way I couldn't believe my eyes. No matter how many photographs you've seen, experiencing it with your own eyes is a unique experience. I think there is as much wonder in the glow of a distant nebula as there is in the spin of a washing machine. It all depends on your perspective. And isn't that just life? Finding meaning, understanding our place in the greatest story of the universe. Look around you: humans built this. Good or bad, humans did it. Nature is not some idyllic paradise, but a place full of risks and dangers, which we overcome, over time, together.

We all came from nothing, and in the end we are nothing, but the space in between is where we make our choice: to live in ignorance and suffering, or to aspire for something better and to dream of utopia.

Because it is so powerful, nuclear energy is the solution to so many of the problems we face. Want to reduce emissions? Build reactors. Need good jobs? Build nuclear fleets. Seeking energy security? Power plants last at least sixty years, compared with twenty-ish years for wind and solar panels. Hoping to increase economic growth while protecting the environment? Build nuclear power plants. Worried about immigration? Developing the world's poorest regions means people like my parents won't have to leave their homes and families to escape energy poverty.

As an aside, throughout my childhood by parents always said that they would move back to India when they retired. But this never happened. I think in some ways their hearts will always belong there, with their extended families, but they haven't even been able to visit the Punjab in over two decades. I suspect that it is too painful for them to see our relatives struggle as they do. There is no going back. Although my parents have done their best to support their families, money alone will not solve the many challenges people face back home.

When we visited my parents' villages in the early 2000s, my parents spent the entire trip telling me not to do things. Don't walk in the long grass, don't wander off alone, don't drink the tap water, and so on. I remember my dad telling me, sternly, that if I heard a pack of dogs, I had to run *immediately.* I replied, 'Dad, didn't you grow up here with these dogs? You said they used to chase you.' He looked sad and didn't respond. Through rose-tinted glasses and the comfort of his quiet home in England, he had forgotten the harsh realities and risks of his childhood in rural India, and now struggled to see his own children in that environment.

There is a common perspective in India that I always found difficult to grasp – that of *kismat*; the idea that if something is going to happen, it will, regardless of one's actions. While hippy westerners embrace this concept as representative of some form of peaceful, free-living Zen, I saw that, in its country of origin, it means rationalising having to live alongside constant defeat and little agency. Relying on the rains to water the crops. Relying on luck to make it through the day. Relying on an imaginary figure, the Almighty, to keep you safe, because the risks of daily life are so great. The only way to make sense of the constant struggle and tragedy is by exporting all agency to the idea of greater beings; the gods.

My parents were religious when they moved to England, but over time they became increasingly secular, like most of the country. There's no need to believe in *kismat* when you have some agency over your own life choices and your fate.

Since I embarked on nuclear advocacy in 2020, the landscape has dramatically changed. For decades, nuclear energy struggled with a reputation burdened by fear, misinformation and high-profile accidents. Back then, a number of prominent environmentalists believed in nuclear power as a sensible and inevitable answer to the climate crisis; but they, including some of my friends in XR, chose not to speak on record out of fear of being silenced.

As I threw myself into challenging misinformation and telling better stories, I also watched an encouraging shift in public opinion take place, with more people than ever in my lifetime supporting nuclear energy as a critical component of a clean energy future.

Although the technical nerds urged me to talk about atoms, I instead talked about energy poverty. I talked about my family, and the way XR had helped to shape language used to describe climate change. I strived to share uplifting stories about clean energy and inspire people to envision a future of energy abundance rather than scarcity.

It began to catch on.

Scientists, engineers and environmentalists who once opposed nuclear energy, gradually began to publicly shift their positions, lending credibility to the technology's safety and necessity. Prominent voices in climate activism and scientific communities now emphasise nuclear energy as a vital tool in reducing global greenhouse gas emissions. New influencers appeared and followed in my footsteps. In 2021, I received an email from a prominent Greenpeace activist who lived and worked on the Rainbow Warrior and had previously been arrested and held overseas for taking part in a major action. When we met up in real life, he expressed frustration with his organisation's continued opposition to nuclear energy, although he wasn't ready to go public about it just yet.

Surveys from around the world reflect this same trend. In the UK, public support has increased for nuclear energy since I entered the arena in 2020. In the US, polls show growing acceptance of nuclear energy, especially among younger generations concerned about climate change. In Europe, countries like Germany and Belgium, that previously shut down nuclear plants, have U-turned. Even in Japan, the narrative is evolving since the Fukushima disaster, with the government re-embracing nuclear power to ensure energy stability.

In 2023, my friend Andreas Fichtner briefly obstructed the destruction of Germany's Grafenrheinfeld Nuclear Power Plant's cooling towers. At that point in time, coal and gas were the largest sources of electricity on the German grid. Fichtner wanted to highlight the absurdity of Germany's policy to phase out nuclear energy altogether, marking a turning point in the world of environmentalism. Although Fichtner was arrested for this first-ever pro-nuclear direct action, it was barely reported by the mainstream press, which has yet to catch up with the changing attitudes to nuclear energy, both within the environmental movement and more broadly. However, I believe history will treat him well.

Moreover, the development of advanced reactor technologies, such as small modular reactors (SMRs) and next-generation designs, has sparked renewed interest in what can be achieved with nuclear technology.

Ultimately, the shift in public perception is a hopeful sign that science and pragmatism are overcoming fear and misinformation. Moreso, the turning of the tide within the environmental movement itself is promising if we want to build a better future. While challenges remain, the increasing support for nuclear energy opens new doors for collaboration and for progress, bringing us closer to a sustainable and resilient energy future. We have found our collective voice, begun to tell new and better stories, and paved the way for industry to take their work forward. In all my years of activism, it is the best work I have done.

CHAPTER FIFTEEN

THE FEMINIST CASE FOR ENERGY ABUNDANCE

'There is no fate but what we make for ourselves.'
Sarah Connor in *Terminator 2: Judgment Day*

I enjoy a good story as much as the next person. The problem, of course, arises when fiction is accepted as fact, and the story then shapes a generation. Take, for example, the prominent anti-nuclear environmentalist, Helen Caldicott, who over her ninety-plus years has persuaded many that nuclear power plants pose the greatest threat to humanity. Caldicott is renowned for her tireless campaigning against nuclear weapons and nuclear energy, and her claims have ranged from asserting that nuclear plants cause cancer to suggesting that nuclear technology is inherently 'male' and therefore aggressive. I am convinced that the fact that more women than men are against nuclear technology is partly due to her rhetoric. If you think I'm exaggerating, here are her exact words at the Occupy Hanford rally held in Richland, Washington, in 2012.[182]

> 'And I'm interested in what causes the nuclear illness affecting our planet... It's these men's brains. And one in twenty-five people are sociopaths, they have no moral conscious, they're brilliant but no moral [conscious]... And we, who are fairly not neurotic and fairly stable, we don't have the energy the neurotic drive to take over the planet and destroy it. So therefore we've gotta rise up. And I would say particularly the women. Because

> we are highly intelligent… We're much more in touch with our intuition and the like… And unless we step up to the plate we're doomed. We're doomed from global warming, we're doomed from nuclear war and we're doomed from nuclear power.'

Similarly, Vandana Shiva spent years demonising gene-editing, and one of her key arguments is that women are 'Knowledge Keepers' – custodians of seed diversity and traditional food knowledge, responsible for preserving the heritage of crops and farming practices across generations. While this perspective highlights the important role women play in agriculture, it also implicitly casts modern biotechnology and scientific innovation as incompatible with women's work, sidelining female scientists and researchers who are actively shaping the future of food security and sustainable farming.

Meanwhile, Caldicott has argued that the development of nuclear technology reflects a patriarchal mindset that prioritises power and control over human life and environmental well-being.

I don't think it's a coincidence that there is a gender disparity regarding support for technologies like nuclear energy. Surveys consistently show that women are generally less supportive of nuclear energy than men, and this gender gap is often attributed to differing levels of risk perception, with women typically expressing greater concern about safety, waste disposal, and environmental impact.

Arguing that women don't belong in science or the energy sector not only perpetuates outdated stereotypes but also erases the remarkable contributions of women who work in and have shaped the field. Pioneers like Maria Salomea Skłodowska-Curie, who discovered radioactivity; Lise Meitner, who played a key role in discovering nuclear fission; Chien-Shiung Wu, whose experiments disproved parity conservation and advanced nuclear physics; and Dorothy Crowfoot Hodgkin, who made critical contributions to crystallography, all demonstrate that women have been central to scientific breakthroughs. Today, countless women nuclear engineers, physicists and technicians continue to advance our understanding of nuclear energy and drive innovation. We *are* the knowledge keepers, just not with the narrow definition Shiva intends.

I would argue that women have benefitted most from new technologies.

So, this one's for the ladies. The terms 'nuke bro' and 'tech bro' are often used as insults in reference to a man who is deemed to be unhealthily in favour of technology. But I am something else – a self-proclaimed 'tech *sis*'. That's because nature is sexist, and technology often balances the scales. Technology has arguably benefitted women far more than it has men; so where are all the tech sisters? Did Helen Caldicott, Vandana Shiva *et al* help to convince a generation of women that technology is the domain of man?

Consider reproductive health. The invention of the birth control pill was a technological breakthrough that gave millions of women control over their fertility and, with it, greater autonomy over their lives, education and careers. In many parts of the world today, apps for period tracking, telemedicine and maternal health monitoring are continuing this legacy, giving women tools to understand and advocate for their own bodies.

Technology has given women freedom in and from motherhood. People often talk of the miracle of childbirth, but the real miracle is the mother surviving pregnancy and delivery. Historically, a significant number of women died from childbirth, with mortality rates as high as 170 per 10,000 births in the mid-17th century and remaining on a plateau in the 19th and early

20th centuries. This was a common and often expected risk, with causes including infection from childbirth, such as childbed fever, and complications from the birth itself. The advent of modern medicine, improved sanitation and hygiene, and a better understanding of childbirth practices led to a dramatic decline in these rates from the mid-20th century onwards. Now, maternal death is extremely uncommon.

Technological advances have given women more choice than they've ever had before. While media focus tends to be on the politics of these choices, the miracle of their existence is often ignored. Yet the development of the pill allows women to enjoy pleasure without risk of pregnancy, and the process of abortion (and all the tools required to undertake it) allow women the freedom if they make a mistake, have a high-risk pregnancy, or are failed by contraceptive measures.

When a woman bleeds, she can now use tampons or sanitary pads to virtually eradicate the outward downsides that, only a few hundred years ago, would have marked her as 'unclean' and made her a social pariah during her monthly cycle. In many cultures throughout history, menstruation meant exclusion from work, education and even family life. Today, the widespread availability of menstrual products has given women greater freedom to move through the world without stigma or restriction. Painkillers, too, help to alleviate the discomfort and physical burden of menstruation, reducing the days lost to pain and fatigue. Together, these innovations are not trivial conveniences but profound enablers of equality, allowing women to participate fully in society in ways their ancestors could scarcely have imagined.

If a woman chooses to have a baby, prenatal screening devices drastically reduce infant mortality rates, epidurals reduce pain during childbirth, and C-sections reduce maternal mortality rates. If a mother cannot breastfeed, she has the option of giving her baby formula milk.

If a woman is unable to have a baby, she now has access to technologies like egg freezing, IVF, and even surrogacy – options that would have been unthinkable for most of human history. Reproductive science has transformed what was once considered fate into an area where choice and possibility exist. Beyond fertility, the story is even bigger: from antidepressants that help manage mental health, to antibiotics that prevent once-deadly infections, women today in much of the world enjoy freedoms and opportunities that their ancestors were categorically denied. Modern medicine and technology don't just extend lifespan – they expand agency, giving women greater control over their own bodies, futures and lives.

The biggest lie women have been told is that technology is somehow at odds with what we stand for. In truth, technology has been our greatest liberator. It has freed us time and time again, and digital platforms have also amplified our voices in ways our grandmothers could scarcely imagine. Far from being an enemy, technology has been the tool that dismantled barriers and expanded our freedom.

As Hans Rosling pointed out in his talk about washing machines, saving his mother from hours of arduous labour meant that both mother and son had the freedom to learn languages and read books. As well as washing clothes by hand, women have historically carried the burden of collecting water. Technology can free areas from water stress – access to clean drinking water at the turn of a tap and desalination for when the rains don't fall. I think of my own mother, escaping a lifetime of toiling over gas stoves in forty-degree heat.

Technology also democratises knowledge. The internet allows women and girls to access learning, especially in regions where formal education is restricted; to find mentors and share their stories. Social media platforms have become spaces where feminist movements mobilise, from #MeToo to grassroots campaigns in countries where speaking out is dangerous. These platforms aren't perfect, but they've made visible what was once silenced.

Technology, as liberation technology, has always been more than machines, gadgets or tools; it has been a means of transformation and freedom. For women, it has been a force for self-definition, an engine for rebellion and a tool of resistance. From the typewriter to the birth control pill, from washing machines to social media, technology has expanded the horizon of what women can do, where they can go and who they can become. It was women, who were dubbed 'human computers', that helped us to land on the moon.

Tech is helping solve problems that disproportionately affect women. Clean cooking stoves reduce indoor air pollution that kills millions – mostly women and children. Mobile banking and digital ID systems are helping women gain financial independence where traditional banking excluded them.

I've spent years arguing that we should wean off of fossil fuels. I still believe we should. I worry about climate change and now see that nuclear energy is environmentally superior.

But I fell for the activist messaging that lumped all fossil fuels together when it's obvious that natural gas and propane, used in the form of liquified petroleum gas (LPG), in South Asia and Africa, are vastly different fuels from coal, wood and dung.

Women and girls are hurt most by the continued use of wood and dung since they are responsible for most of the cooking in underdeveloped nations. Both through exposure to fumes through cooking, and by missing out on education to spend their time collecting wood or cow dung for fuel.

Yet for a long time wood and dung have been considered

'renewable' while propane has been maligned as the dirtier fuel. In truth, greenhouse gas emissions per unit of electricity generated from biomass is often higher than those from gas.

Replacing coal with natural gas is the main reason the US, UK and dozens of other nations have significantly reduced their carbon emissions in recent decades. Phasing out gas is misguided and will harm people in low-income countries, where LPG is essential for survival. Consider India and Brazil: during the energy crisis, they could not afford enough gas to fully power their economies, which led to blackouts in Bangladesh and Pakistan. Instead, LPG imports went to richer countries that could afford them. This underscores the importance of countries using their domestic resources rather than taking imports destined for poorer nations and relying on exports, as Germany has done since shutting down its nuclear power plants.

Billions of poor people in the world are striving for energy equality. As much as I'd like them to have nuclear power plants, for many countries this is an unrealistic immediate goal; historically, such leapfrogging does not work for poorer countries. In the meantime, wind and solar power alone will not enable poorer nations to achieve the growth and development they need to achieve a high quality of life; it is not even working in the West.[183] Still, unlike them, we can rely on importing expensive energy when wind and solar generation are low. They will need fossil fuels as a cheap and cleaner option.[184]

When my parents migrated to England they moved to Birmingham, known then as the 'city of a thousand trades'; it was going through an industrial boom, with factories making everything from cars through pots and pans to electrical goods.

As an environmentalist this appreciation for 'consumerism'

always posed an obvious dilemma. I have advocated for living with less, and making objects as sustainably as possible, but equally I recognise that people need pots and pans and that the more eco-friendly items often cost more (and are not always better for the environment). I know that we need to bring down greenhouse gas emissions, but my efforts have now shifted away from expecting ordinary people to shoulder this burden or to make drastic lifestyle changes.

Manufacturing jobs were once the lifeblood of the British economy. Birmingham's industry flourished in the three decades that followed the end of the Second World War; in fact it thrived, even above other cities in Britain at the time, thanks to a combination of immigration, innovation and civic pride.

Growing up, I experienced what reliable jobs offered families like mine, and later I saw the heart of British industry stripped away as those jobs were increasingly outsourced abroad. As migrants, these jobs had given my parents more than a higher quality of life and a better future for their children; they also gave them a sense of belonging. Community. Identity. Pride.

My older brother, who would play with Lego with me for hours and who never seemed to tire of explaining the inner workings of, say, a fridge to a curious ten year old, did not go to university when he turned eighteen. My parents were adamant that their children work instead (a trend I bucked, but that is another story). My brother took an apprenticeship to become an engineer, and worked his way up the ladder. Years later he contributed to the design and building of some of Birmingham's bigger landmarks, including the large university hospital and the redesign of the town centre. Can a child of immigrants contribute more significantly to their home than by building elements of the country with their own hands?

Then the recession began, in 1980, with 20,000 redundancies

a week announced during a single month, all in manufacturing. As those opportunities disappeared, unrest grew. In 1968, the Wolverhampton MP, Enoch Powell, delivered his notorious Rivers of Blood speech, addressing disgruntled Brummies and putting the blame squarely on many of the pot makers and engineers. I had experienced the multicultural bounty of my city: as a child I'd watched Aston Villa football matches (we lived in the inner city, near the stadium) and looked forward to the large gatherings celebrating Bonfire night, but I'd also anticipated the yearly *mela* (festival) in the park at Diwali. Hot dogs and candy floss were eaten at one event and samosas and ladoos at the other. These events were not segregated, and there was room for both.

I watched the balance erode. The collapse of Birmingham's industrial economy was catastrophic. In 1976, the West Midlands region still had the highest GDP of any in the UK outside the South East, but within five years it was lowest in England. A foundation of Birmingham unravelled.

How much of the current state of affairs in Britain is due to this loss of industry? Economic downturn often leads to social and political unrest. In Britain at present, people are struggling to make ends meet and many parents can't afford to feed their children even when working full time. Living wages are low. Energy is expensive.

A difficult pill to swallow is that sustainable lifestyle changes are hardest for working-class people. Owning an electric car or solar panels, taking the train abroad rather than a much cheaper flight, and buying quinoa instead of *keema* (an Indian dish made of ground meat) are predominantly middle-class choices that require adequate income and time. I have long argued that it should be cheaper for people to make green choices, for example, with more low-cost options for public transport, but what working-class communities need first and foremost is opportunities.

We faced this same problem in XR, where we struggled to appeal to working-class and diverse communities. Researchers found that the movement was 'largely white, middle-class and highly educated'.[185] This topic raged as a debate for the entire time XR was active, and I would try to make the points I have outlined here, but to no avail. After leaving the organisation, in 2022 I went head to head with co-founder Clare Farrell on the *Sky News* programme, *Common Ground*, hosted by Trevor McDonald,[186] where we tried and failed to see each other's perspectives. The trouble was that I *did* see hers, having been a core member of the group myself and an environmentalist for much longer; but she couldn't see mine. She said she didn't understand why more people weren't out protesting about climate change. I said, 'Because people are worried about putting food on the table… trying to feed their families.' But there was no common ground.

And this was no mistake. As one of the few people of colour in the movement, XR founders often asked me to take speaking slots just to address the diversity issue – something I wasn't keen to do, but generally accepted in my role of spokesperson. But that was all for show and as far as their concern really stretched. In October 2019, XR activists staged a rush-hour protest at Canning Town station in East London. They climbed onto the roof of a Jubilee Line train, blocking services during a busy commute. The action was part of a broader strategy to disrupt public transport in three locations: Canning Town, Stratford and Shadwell.

As soon as I heard about the action, I spoke out against it. Most of the co-founders, however, supported it, telling me, 'Disruption is what we do.' I knew that, and the issue for me wasn't the tactic of non-violent direct action itself, but *where* they were choosing to carry it out. These were underfunded, predominantly low-income areas. Why not stage actions in wealthier parts of London instead? I put this question to one of the organisers directly. He admitted they had already scoped out those areas, but found the security too tight to execute their plans. So, they had *deliberately* chosen poorer areas, where security was lighter and resistance weaker, because those areas were underfunded. That decision stayed with me, as it revealed whose lives they were willing to disrupt, and whose they weren't.

When the actions took place, there was immediate chaos. Commuters who were delayed became visibly angry. Videos showed passengers being chased, pulling protesters off the train and physically assaulting them – some punching and kicking them as they lay on the ground.

I watched the footage with deep sadness and discomfort. A woman was pleading with one of the activists to let the train move or she was going to be late for work and feared losing her job. In that moment, it was painfully clear to me that poverty, racial tensions and the realities faced by the working-class weren't priorities for this group, or for any environmental organisations I had previously been involved with. We weren't fighting for the same thing. To them, people were the problem – so their suffering could be dismissed in service of a greater ecological goal. For me, the climate fight has always been about people. The challenge of solving climate change is not simply to protect nature, but to secure a better, fairer future for humanity. After all, the world will continue to spin whether or not we are on it.

And that's why we could never find common ground.

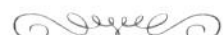

When I visited Suffolk in southeast England to visit the proposed site for a new nuclear power station, Sizewell C, I asked people what they thought of the idea. Nine out of ten locals I spoke to were in favour of the plant and all of them mentioned jobs and training, especially opportunities for their young people who will, without said jobs, move away. Hence the unions Unite, GMB, and Prospect support the construction of Sizewell C.

At first I was amused to see that local residents had graffitied on anti-Sizewell posters that had been erected by anti-nuclear activists around Leiston, but then it made me sad. This activism is drawn clearly along class lines. It makes me uncomfortable to see opposition to well-paid job opportunities from celebrated individuals like actor Bill Nighy, who owns a second home in Suffolk, and ex-broadcaster Bill Turnbull who also campaigns against Sizewell C. It is telling that Turnbull does not mention jobs in his commentary against the nuclear power plant.

This sort of activism comes from a privileged position from those with the means, time, celebrity or public platform to share their views. Many ordinary workers lack the time and ability to do the same, yet they are making just as significant a contribution to society. Heads down, working with their hands, they are getting on with the job.

The idea implied in climate activism, that working-class people don't care about the planet while those with electric cars and solar panels do, is unfair and incorrect. Many of the workers who choose to take jobs at nuclear plants do so because they care about the planet and their children's futures, just as my parents and grandparents cared about mine. It just doesn't make headlines.

Workers in the nuclear industry also experience stigma, from mild mockery engendered by *The Simpsons* caricature to outright enmity in Germany from the anti-nuclear movement, even though they are doing immense good in the world, and they do so without glory. Yes, they want jobs, but they want good jobs too, and jobs that make our world a better place.

For me, new nuclear plants provide a tidy outcome: I want to see a return of community to Britain's streets, for hardworking people to enjoy candy floss and samosas without increasingly fighting among themselves. I want to advocate for sustainable work options, to tackle air pollution and climate change, and for people to have good employment opportunities. Building nuclear power plants ticks all these boxes nicely.

Politicians and celebrities can debate climate policies until the cows come home, but the reality is that working-class labourers will deliver what Britain needs to tackle air pollution, climate change and address slow growth. Yes, those in power need to sign the documents, but the hands that built these cities are the ones that will save the world.

CHAPTER SIXTEEN

THE FUTURE IS AN UNWRITTEN STORY

'All we have to decide is what to do with the time that is given to us.'
Gandalf, *The Lord of the Rings*

It's amazing how quickly things change.

Just six years ago, speaking favourably about nuclear energy was frowned upon both in public and in private. I lost count of the number of times I was called a shill for doing this work, whether for 'Big Pharma' when discussing gene-editing, or 'Big Nuclear' when discussing clean energy. The truth is that I *am* a shill. I am a shill for the human race. I want to see the end of needless suffering and death from preventable diseases. I want to see all poverty eradicated. Apart from a few specific war-mongers, who doesn't want to see the end of all war? I want life on this planet to thrive. I want to see the best of humankind during my brief presence on this pale blue dot that we are fortunate enough to inhabit. I want us to live long and prosper.

Humans are capable of solving immense problems, and achieving great things through forging strong values and working together. I am a shill for progress because I want to see what we do next, once the immediate and entirely solvable problems, like air pollution, poverty and climate change, have been solved.

I am going to end where we started: by talking about the other timeline. I may not be in that other timeline, bending over a smoky stove, but my family members still are. In my parents' village, a girl the same age as my youngest daughter has already learned how to gather and split wood for cooking. She already has that deep, rasping cough common among those who every day breathe in smoke from these stoves. She walks barefoot and she prays for the rains to come. I hope she doesn't hurt herself while bending over the stove. In this part of the world one bad injury could alter the course of her life, and not for the better.

Development will change their lives. Gas stoves will liberate these children and women from some debilitating health impacts. Connecting to the grid and gaining access to electricity will liberate them from virtually all their burdens; from air conditioning, protecting them from excessive heat, to electric or gas cookers. Access to plumbing brings clean water. Electric cooking brings cleaner air. Gone is living at the mercy of the weather. Now let's add further development – desalination, roads, schools, phones with AI, so they can access the breadth of human knowledge even without being literate.

But let's be honest: what will most liberate them is wealth. Personal wealth and freedom, development to grant them access to clean water, energy, shelter; these are not luxuries, but human rights. This wealth does exist in India. It's just not evenly distributed, and opportunities are absent if you are born in the wrong part of the country. This is the tale of poverty the world over.

For most of us, 'poverty' is an abstract concept, and that's ultimately a good thing. I dream of a world where there is no concept of poverty.

Back in my parents' home village in India, the journey from the airport to the village takes over four hours by car, mostly on mud tracks and sometimes through flooding. During monsoon season I've watched people wading through waist-high water with children on their backs, or carrying bags above their heads. Women in saris making the trek from their village to the airport. I've wondered how long it would take them to get there, and whether they'd get sick from trudging in floodwater. I wondered about the snakes. I wondered why it was so important for them to risk those journeys.

I spent three full weeks in India wondering similar things.

When we finally arrived at the village I was greeted with tearful, joyful faces and grasping, calloused hands. We were like celebrities to the village: children of the family that had escaped; children who had finally returned. Yet, looking around at the way my relatives were forced to live, I felt a deep and lasting sadness.

Everywhere I looked in India I saw intense beauty, but it was always the backdrop to signs of poverty. Vibrant, beautiful landscape, and stories of cousins we had lost to snake or dog bites. Long family trees, engagements and weddings, and missing family members who had never returned home (probably due to car accidents). Imagine a world without your phone, laptop, GPS, car, cooker, flush toilet, lighting after dark, and so on, and then add the risk of a hundred different things that are out of your control. Then apply that risk to your loved ones. This is how billions of people still live today.

Their poverty is technology poverty.

I am generalising based on my mother's village. As I say, there is wealth in India. And villages have different stages of development; some homes have brick walls, small houses and temples. If the village has a motorbike, that's a boon for everyone. And of course, energy poverty isn't just a problem in faraway places – it exists here in Britain too. Electricity is expensive, and that cost takes a toll. Families are forced to choose between heating and eating. Every winter, people die because they cannot afford to keep their homes warm. In a wealthy country, this is not just tragic, but utterly unacceptable. Reliable, affordable energy isn't a luxury. It's a matter of survival.

No matter where you are in the world, then the equality barrier is crossed with access to energy.

I watched women cook over small stoves, burning rudimentary fuels like wood or charcoal that left them coughing and wheezing, and I was alarmed whenever children ran over to help out. Always girls. No one I met had regular access to electricity, though one wealthier family had a generator to use some of the time and considered it a luxury. I felt helpless and somewhat guilty about my own way of life back home; the sheer abundance of power – figuratively as well as literally; and about the scale of the problems people face in these villages, without the development we reap the benefits of at home.

In joining with environmental groups that told me the problem was capitalism, and the solution was degrowth, I was muddled and misled for many years. Like my eco-peers, I initially spent time feeling guilty, and self-flagellating by making personal sacrifices and advocating for living more simply. But I stepped away from the movement when I realised I had been sold a lie, and the inverse is true: we are fortunate, but we don't have to relinquish our good fortune. It doesn't benefit anyone. We can fight for others to attain a high quality of life as well. What has benefitted us at home, and what countries like India need, is *more* energy, not less.

We can't doom the world's least resilient countries to the consequences of climate change. To help slow climate change, that energy needs to be clean, and there is no real excuse for developed nations not to pivot to clean energy immediately.

I got to know my family in the Punjab and met the cousin who had fits as a baby and was severely disabled as a result. My parents – wealthy now by villager standards – had tried to source a wheelchair for him, and failed. There is no Amazon Prime in the village. I watched as he was carried from chair to bathroom to bed by human arms. The saddest thing about this was that I knew a close friend back home in England who had a similar ordeal when his daughter was born, yet she now leads a relatively normal life thanks to easy access to medication that quickly stopped her fits.

People who live in poverty just want the chance to have a better life. This drive is what compelled my parents to board a boat to Britain, a country they couldn't (then) point to on a map, where people spoke in an unfamiliar tongue, to leave behind everything they knew, to offer their children a better life.

My parents will never retire to India. The fact is that they can't. Stories of rabid dogs, fateful accidents, untreated illness and fatal snake bites shocked them as much as they shocked me. (India has hospitals with anti-venom, but the nearest was hours away, with minimal village transport to get you there. If you wonder why the culture reveres doctors, this offers insight). That's the only problem with experiencing a high quality of life – you can never go back.

The future I imagine is not one of scarcity and sacrifice, but of abundance, dignity and possibility. A future where clean air, reliable electricity and access to modern medicine are not luxuries, but birthrights. Where the promise of technology is not withheld in the name of ideals born from comfort, but embraced as a tool of liberation.

It's taken me years to unlearn the romantic myths I absorbed from energy-rich societies – myths that equate simplicity with virtue, and deprivation with authenticity. But poverty is not noble, or a culture to be preserved. It is a condition to be overcome.

The people I've met and written about throughout this book don't need to be saved. What they need is the freedom to choose their own path; to harness the full spectrum of technologies that the world has to offer, just as wealthy nations did before them. That means building infrastructure, nuclear reactors, desalination plants, cooling systems and digital networks. Building futures.

More, nuclear energy is a crucial part of the solution to myriad environmental challenges, including climate change. Unlike fossil fuels, nuclear power generates minimal greenhouse gas emissions and air pollution, while helping to decarbonise energy systems while maintaining the constant supply needed to support modern life. Its high energy density also means it requires far less land than other energy sources, preserving natural habitats and reducing environmental disruption. Nuclear can stabilise grids, reduce reliance on coal and gas, and accelerate the transition to a sustainable, climate-resilient future.

I've spoken a great deal about India in this book, largely because it is the country I have ties to, though I would not claim to know it best as there is a cultural divide that cannot be surpassed. While my experience, research and personal connections have rooted my arguments there, they are not unique to India. They apply equally to many other low- and middle-income countries across Africa, Southeast Asia and Latin America; places where energy poverty remains a daily barrier to health, opportunity and human dignity. Whether it's Nigeria, Bangladesh, Kenya or Bolivia, the core issue is the same: access to abundant, reliable energy is the foundation on which education, healthcare, gender equity and climate resilience can be built. These countries, too, deserve the full range of technological options, including clean, abundant energy, to chart their own futures.

They don't want handouts. They only need not be held back. Don't patronise them by offering a few solar panels for the village (as Western NGOs do). Just step out of the way when they say they want to build power plants (which Western nations ask them not to do because of climate change). With economic and energy growth, they will also gain access to clean water, modern technologies, reliable healthcare and education, all of which are essential for improving quality of life and enabling sustainable development.

We cannot claim to care about climate justice while denying

billions the tools they need to thrive in a warming world. We cannot speak of equality while urging restraint from those who have never had the chance to indulge. And we cannot say we care about the environment while simultaneously protesting and telling mistruths about one of the cleanest, safest and most powerful forms of energy.

I believe a better world is within reach. Not by returning to the past, but by stepping bravely into the future. One powered not only by clean energy, but by compassion, reason and the fierce conviction that every human life deserves light, warmth and hope.

To me, that is what saving the planet truly means; not preserving it in isolation, but protecting it for the life that lives on it. That is where my journey has led me, and is the path I remain committed to, however uncomfortable this story might be to digest. I will continue to fight for the change we need and stand with the voices that do not get heard. Call me a shill, call me a fraud; I can bear it because I am one of the luckiest people in the world, to have given my children an energy-rich life. When I go to bed tonight, I know our home will be warm. When I wake up in the morning, I know I can cook breakfast without choking on smoke. I want that simple security and dignity for every person on this planet.

And that is as close as I'll ever get to calling anything *kismat.*

AUTHOR'S NOTE

In this book I often use the words *energy* and *electricity* interchangeably. Strictly speaking, they are not the same thing: energy can take many forms (heat, chemical, kinetic, nuclear), while electricity is just one way of delivering and using it. But because electricity is the most visible and transformative form of energy in modern life – powering our lights, heating, communications and increasingly our transport – most readers intuitively understand 'energy' in terms of electricity. So I have used the terms interchangeably to keep the narrative accessible without losing the core message.

ACKNOWLEDGEMENTS

Writing this book has been a journey of curiosity, collaboration and reflection, and I owe immense thanks to so many people who have walked alongside me. It would be impossible to name everyone, but I must especially thank John Stumbles for his endless patience and encouragement, and the pro-nuclear advocates who inspired me to tell stories that matter and stood with me when times were tough, including Myrto Tripathi and everyone at *Voices for Nuclear* in France, Andreas Fichtner, Robert B. Hayes, James Hansen, Rod Adams, Wade Allison, Susie Ho, Michelle Robichaud, Megan W. Owen, Graham Young, Atte Harjanne, Patrick A. Burr, Meredith Angwin, and so many others.

To the *Emergency Reactor* members who spent countless hours myth-busting with me and handing out bananas: Arun Khuttan, Rob Loveday, Mark Dawes, Alex Clements, Andrew Alan Key, Dominic Adams, Jess Cliff, Martin Donnohue, Marie Zabell, John Stewart, Robin Wilde, Owen Thomas, Alasdair Lumsden, and Trevor Rix – who sadly passed away in 2025 – thank you for your passion and solidarity.

I am also deeply grateful to my friends and family for their support, especially Sophie Roberts and David Wilson.

I owe thanks as well to the many readers, viewers and listeners who have engaged with my science communication over the years; your curiosity and enthusiasm have kept me going. To the interviewers who listened with an open mind and were willing to change their views on nuclear – you helped to make a difference. Special thanks also to Claude Roessiger for supporting my work, Humanists UK for providing a community where reason, evidence, and human progress are celebrated, Liz Sweet, Sabine Hossenfelder, Daniel Aegerter and family and Rune.

To the publishing team at Unicorn, thank you for believing in this book.

Finally, to anyone who has changed their mind: thank you for demonstrating that the freedom to rethink is humanity's greatest strength.

I hope this book has inspired you to ask bold questions, embrace evidence, and imagine a future where energy, innovation, and human ingenuity come together for the common good.

Live long, and prosper.

ENDNOTES

[1] https://abcnews.go.com/Politics/wright-increase-domestic-energy-production-dirty-energy-clean/story?id=117721233

[2] Mark Andreessen in his essay The Techno-Optimist Manifesto, 2023. https://a16z.com/the-techno-optimist-manifesto/

[3] https://www.cutcarbonnotforests.org/scientist-letter-read/

[4] https://www.britannica.com/event/Typhoon-Nina-Banqiao-dam-failure

[5] https://www.bloomberg.com/news/articles/2023-01-16/cost-of-managing-germany-s-power-grid-jumped-to-record-in-2022?embedded-checkout=true

[6] https://ec.europa.eu/eurostat/web/products-eurostat-news/-/ddn-20201124-1

[7] When Ukraine became independent, the government requested that people re-establish the original spellings of Ukrainian places. This article therefore uses their preferred spelling 'Chornobyl' over the Russian spelling 'Chernobyl'. It is not a typo. https://www.cbc.ca/news/canada/new-brunswick/chornobyl-spelling-editors-note-1.3765167

[8] https://www.ft.com/content/967e1d77-8d3c-4256-9339-6ea7025cd5d3

[9] https://pubs.acs.org/doi/10.1021/es3051197#:~:text= Using%20historical%20production%20data%2C%20we,resulted%20from%20fossil%20fuel%20burning.

[10] https://news.climate.columbia.edu/2020/01/27/economic-growth-environmental-sustainability/

[11] https://www.jstor.org/stable/3744148?typeAccessWorkflow =login

[12] https://blog.greenprojectmanagement.org/index.php/2019/05/13/pollution-why-we-replaced-horses-with-automobiles/

[13] https://www.who.int/news-room/fact-sheets/detail/household-air-pollution-and-health

[14] https://www.ceh.ac.uk/news-and-media/news/uk-air-quality-improvements-over-40-years-cut-death-rates

[15] https://pubmed.ncbi.nlm.nih.gov/10845777/

[16] https://www.epa.gov/sites/production/files/2020-05/2019_baby_graphic _1970.png

[17] https://www.aeaweb.org/articles?id=10.1257/aer.20151272
[18] https://ember-energy.org/latest-insights/uk-biomass-emits-more-co2-than-coal
[19] https://www.theguardian.com/business/2023/feb/23/drax-power-station-profits-nearly-double-call-for-subsidies-cut
[20] https://www.bloomberg.com/news/articles/2023-10-16/germany-fires-up-reserve-coal-plant-in-winter-s-first-cold-snap
[21] https://haas.berkeley.edu/wp-content/uploads/WP304.pdf
[22] https://japannews.yomiuri.co.jp/business/companies/ 20230210-90175/
[23] https://www.investopedia.com/terms/t/thomas-malthus.asp
[24] https://ourworldindata.org/grapher/agricultural-land
[25] https://papers.ssrn.com/sol3/papers.cfm?abstract_id=4535022
[26] http://www.pnas.org/content/104/17/7301.long
[27] https://www.youtube.com/watch?v=5JiYcV_mg6A
[28] https://www.weforum.org/agenda/2023/09/elderly-oldest-population-world-japan/?ref=hir.harvard.edu
[29] https://www.theregister.com/2024/05/15/japan_electricity/
[30] https://ourworldindata.org/explorers co2?tab=map&facet=none&country= CHN~USA~IND~GBR~OWID_WRL&Gas=CO%E2%82%82&Accounting=Consumption-based&Fuel=Total&Count=Per+country
[31] https://thebreakthrough.org/issues/energy/absolute-decoupling-of-economic-growth-and-emissions-in-32-countries
[32] https://publications.parliament.uk/pa/cm201213/cmselect/cmenergy /117/117we19.htm
[33] https://www.bbc.co.uk/news/uk-54103163
[34] https://www.cancer.gov/about-cancer/causes-prevention/risk/substances/vinyl-chloride
[35] https://www.bts.gov/content/train-fatalities-injuries-and-accidents-type-accidenta
[36] https://www.world-nuclear.org/information-library/nuclear-fuel-cycle/nuclear-power-reactors/appendices/rbmk-reactors.aspx
[37] Reaktor Bolshoy Moshchnosti Kanalny, meaning ‘high-power channel reactor’
[38] https://www.iaea.org/newscenter/focus/chernobyl/faqs
[39] https://cancerletter.com/articles/20190524_3/
[40] https://citeseerx.ist.psu.edu/viewdoc/

download?doi=10.1.1.176.254 &rep=rep1&type=pdf

[41] https://www.nrc.gov/docs/ML1234/ML12340A564.pdf

[42] http://heal.tttp.eu/sites/all/modules/civicrm/extern/url.php?u=30667&qid=506802

[43] https://haas.berkeley.edu/wp-content/uploads/WP304.pdf

[44] https://www.sciencedirect.com/science/article/pii/S0301421519303611

[45] https://www.unscear.org/docs/publications/2015/UNSCEAR_WP_2015.pdf

[46] https://www.who.int/news-room/q-a-detail/health-consequences-of-fukushima-nuclear-accident

[47] https://www.forbes.com/sites/jamesconca/2018/09/06/no-the-cancer-death-was-probably-not-from-fukushima/

[48] https://www.nrc.gov/reading-rm/doc-collections/fact-sheets/3mile-isle.html#effects

[49] https://www.energy.gov/ne/articles/5-facts-know-about-three-mile-island

[50] https://www.iaea.org/newscenter/focus/chernobyl/faqs

[51] https://www.spf.org/en/global-data/book_fukushima.pdf

[52] https://www.edweek.org/leadership/tv-brought-the-trauma-to-classroom-millions/1986/02

[53] https://www.reaganlibrary.gov/archives/speech/address-nation-explosion-space-shuttle-challenger

[54] https://www.ncbi.nlm.nih.gov/pmc/articles/PMC5763351/

[55] https://www.nature.com/articles/s41467-017-02036-8

[56] https://www.whitehousehistory.org/the-white-house-gets-electric-lighting

[57] https://www.ncbi.nlm.nih.gov/pmc/articles/PMC4924574/

[58] https://www.bbc.co.uk/news/world-asia-54658379

[59] https://theconversation.com/no-the-fukushima-water-release-is-not-going-to-kill-the-pacific-ocean-200902

[60] https://doi.org/10.1155/2020/4834965

[61] https://www.ncbi.nlm.nih.gov/pmc/articles/PMC7770185

[62] https://pubmed.ncbi.nlm.nih.gov/17320504/

[63] https://www.intechopen.com/chapters/57180

[64] https://injuryfacts.nsc.org/all-injuries/preventable-death-overview/odds-of-dying/

[65] https://www.cdc.gov/nceh/radiation/smoking.htm

[66] https://britishskydiving.org/how-safe/

[67] https://theconversation.com/no-the-fukushima-water-release-is-not-going-to-kill-the-pacific-ocean-200902

[68] https://www.cornwall.gov.uk/environment/environmental-protection/radon/ Working in a nuclear power plant typically exposes a worker to around 0.18 mSv of radiation a year. Taking a transatlantic flight exposes you to around 0.08 mSv. These are still small, unproblematic amounts of radiation that do not pose harm to human health.

[69] https://physicsworld.com/a/radioactive-decay-accounts-for-half-of-earths-heat/

[70] https://www.bmj.com/rapid-response/2011/10/28/radioactivity-cigarettes%20https://pubmed.ncbi.nlm.nih.gov/19718986/

[71] http://www.ncbi.nlm.nih.gov/pubmed/7088089

[72] https://www.nasa.gov/analogs/nsrl/why-space-radiation-matters

[73] https://www.scientificamerican.com/article/coal-ash-is-more-radioactive-than-nuclear-waste/

[74] https://www.snmmi.org/ClinicalPractice/content.aspx

[75] https://ourworldindata.org/nuclear-energy

[76] https://www.hsph.harvard.edu/news/hsph-in-the-news/reliance-on-coal-linked-with-lung-cancer-incidence/

[77] https://www.sciencedirect.com/science/article/pii/S2666759220300500

[78] https://world-nuclear.org/information-library/safety-and-security/radiation-and-health/naturally-occurring-radioactive-materials-norm.aspx

[79] https://www.nei.org/fundamentals/nuclear-waste

[80] https://www.nei.org/fundamentals/nuclear-waste

[81] https://www.youtube.com/watch?v=UOVNTcc-vPw&t=467s&ab_channel=TrainsTrainsTrains

[82] Cask test video from 1978: https://www.youtube.com/watch?v=Bu1YFshFuI4

[83] https://www.iaea.org/topics/nuclear-safety-conventions/joint-convention-safety-spent-fuel-management-and-safety-radioactive-waste

[84] https://www.cnbc.com/2022/06/02/nuclear-waste-us-could-power-the-us-for-100-years.html

[85] https://www.hsph.harvard.edu/c-change/news/fossil-fuel-air-

pollution-responsible-for-1-in-5-deaths-worldwide/
[86] https://www.hsph.harvard.edu/c-change/news/fossil-fuel-air-pollution-responsible-for-1-in-5-deaths-worldwide/
[87] https://www.researchgate.net/publication/342671383_Metal_dissolution_from_end-of-life_solar_photovoltaics_in_real_landfill_leachate_versus_synthetic_solutions_One year_study
[88] https://www.bloomberg.com/news/features/2020-02-05/wind-turbine-blades-can-t-be-recycled-so-they-re-piling-up-in-landfills
[89] https://unece.org/sites/default/files/2022-04/LCA 3_FINAL%20March%202022.pdf
[90] https://www.cnbc.com/2022/06/02/nuclear-waste-us-could-power-the-us-for-100-years.html
[91] https://www.iaea.org/nuclear-safety-and-security-in-ukraine
[92] https://www.theguardian.com/world/ng-interactive/2020/jun/01/climate-protesters-german-village-coal-still-king
[93] https://odi.org/en/about/our-work/climate-and-sustainability/faq-coal-poverty-and-the-environment/
[94] https://hummedia.manchester.ac.uk/institutes/gdi/publications/workingpapers/futuredams/futuredams-working-paper-004-hay-skinner-notron.pdf
[95] https://www.repository.law.indiana.edu/cgi/viewcontent.cgi?article=1615&context=ijgls
[96] https://www.sciencedirect.com/science/article/pii/S0303243421002312
[97] https://www.theguardian.com/global-development/2019/mar/11/fires-of-jharia-spell-death-and-disease-for-villagers-india-coal-industry
[98] https://www.ega.edu/facweb2/stracher/2004%20Stracher%20&%20Taylor.pdf
[99] https://www.reuters.com/world/middle-east/turkeys-unfinished-akkuyu-nuclear-plant-not-damaged-by-quake-rosatom-official-2023-02-06/
[100] https://www.audubon.org/news/more-one-million-birds-died-during-deepwater-horizon-disaster
[101] https://www.nytimes.com/2010/11/06/science/earth/06coral.html
[102] https://www.eurekalert.org/news-releases/566817
[103] https://www.nwf.org/magazines/

[104] https://www.thenation.com/article/archive/investigation-two-years-after-bp-spill-hidden-health-crisis-festers/
[105] https://www.britannica.com/list/9-of-the-biggest-oil-spills-in-history
[106] https://upgrader.gapminder.org/t/2017-gapminder-test
[107] https://ozone.unep.org/treaties/montreal-protocol
[108] https://www.nature.com/articles/s41586-021-03737-3
[109] https://ourworldindata.org/extreme-poverty
[110] https://pmc.ncbi.nlm.nih.gov/articles/PMC10787413/
[111] https://www.port.ac.uk/news-events-and-blogs/news/breakthrough-in-enzymatic-plastic-recycling-cuts-costs-and-emissions
[112] https://www.gosh.nhs.uk/news/revolutionary-eye-gene-therapy-gives-children-life-changing-improvements/
[113] https://pmc.ncbi.nlm.nih.gov/articles/PMC12305139/
[114] The Ultimate Guide to Green Parenting, Zion Lights, 2015.
[115] https://www.ncbi.nlm.nih.gov/pmc/articles/PMC3068720/
[116] https://www.salon.com/2017/08/24/gwyneth-paltrow-goop-truth-in-advertising/
[117] https://www.gq-magazine.co.uk/lifestyle/article/coronavirus-5g-conspiracy
[118] https://www.forbes.com/sites/callumbooth/2024/10/16/the-internet-is-angry-about-russell-brands-magical-amulet-heres-why/
[119] https://www.europarl.europa.eu/RegData/etudes/BRIE/2023/747107/EPRS_BRI(2023)747107_EN.pdf
[120] https://www.sciencealert.com/this-mysterious-fire-in-australia-has-been-burning-non-stop-for-at-least-6-000-years
[121] https://ourworldindata.org/child-mortality
[122] https://doi.org/10.1289/ehp6407
[123] https://upgrader.gapminder.org/q/21/explanation
[124] https://www.washingtonpost.com/climate-solutions/2024/05/11/cold-water-laundry-shower-dishes/
[125] https://www.nytimes.com/2020/02/03/climate/japan-coal-fukushima.html
[126] https://www.livescience.com/chernobyl-exclusion-zone
[127] https://news.uga.edu/animal-life-thriving-around-fukushima/
[128] https://www.youtube.com/watch?v=iXAsirlbFFw
[129] https://www.nasa.gov/feature/warming-makes-droughts-extreme-wet-events-more-frequent-intense

[130] https://www.sciencedirect.com/science/article/abs/pii/S0309170816 302044
[131] https://classics.mit.edu/Aristotle/meteorology.2.ii.html
[132] https://www.azcentral.com/story/news/local/arizona-environment/2019/11/29/middle-east-oman-water-desalination-reliance-costs/2123698001/
[133] https://www.ceicdata.com/en/israel/environmental-water-made-available-for-use-oecd-member-annual/water-exported
[134] https://subscriber.politicopro.com/article/eenews/2023/10/06/new-orleans-water-good-news-leaves-regionwide-questions-00120376
[135] https://www.tandfonline.com/doi/full/10.1080/07900627.2022.2138135#d1e135
[136] https://www.tandfonline.com/doi/full/10.1080/07900627.2022.2138135
[137] https://www.vox.com/2015/3/23/8278085/singapore-lee-kuan-yew-air-conditioning
[138] https://www.nature.com/articles/d41586-023-02934-6
[139] https://www.gao.gov/products/gao-24-106213
[140] https://ourworldindata.org/weather-forecasts
[141] https://documents1.worldbank.org/curated/en/896971468194972881/pdf/102725-PUB-Replacement-PUBLIC.pdf
[142] https://www.npr.org/sections/goatsandsoda/2016/12/09/504540392/dial-m-for-money-can-mobile-banking-lift-people-out-of-poverty
[143] https://www.forbes.com/sites/danielrunde/2015/08/12/m-pesa-and-the-rise-of-the-global-mobile-money-market/?sh=17b9e0245aec
[144] https://dictionary.cambridge.org/us/dictionary/english/technology
[145] https://blogs.scientificamerican.com/guest-blog/natures-nuclear-reactors-the-2-billion-year-old-natural-fission-reactors-in-gabon-western-africa/
[146] https://uk.sports.yahoo.com/news/anti-glasses-influencer-samantha-lotus-174507703.html
[147] https://www.thetimes.co.uk/article/old-enemies-in-texas-unite-against-nuclear-dump-czslqnlw9
[148] https://www.theguardian.com/environment/2014/jun/19/russia-secretly-working-with-environmentalists-to-oppose-fracking
[149] https://a16z.com/the-techno-optimist-manifesto/

[150] https://www.scientificamerican.com/article/we-are-living-in-a-climate-emergency-and-were-going-to-say-so/
[151] https://jcom.sissa.it/article/pubid/JCOM_1906_2020_C01/
[152] https://www.thelancet.com/journals/lanplh/article/PIIS2542-5196(21)00278-3/fulltext
[153] https://www.humanprogress.org/what-might-an-energy-rich-future-look-like/
[154] https://www.eia.gov/todayinenergy/detail.php?id=36692
[155] https://www.washingtonpost.com/world/2019/06/28/europes-record-heatwave-is-changing-stubborn-minds-about-value-air-conditioning/?noredirect=on
[156] https://www.sciencedirect.com/science/article/pii/S0301421516300106
[157] https://www.forbes.com/sites/jamesconca/2015/10/22/china-shows-how-to-build-nuclear-reactors-fast-and-cheap/?sh=59d041ef5484
[158] https://infrastructure.planninginspectorate.gov.uk/wp-content/ipc/uploads/projects/EN010012/EN010012-001779-SZC_Bk6_ES_6.1_Non_Technical_Summary.pdf
[159] https://www.eadt.co.uk/news/23260722.suffolk-work-start-sizewell-c-marsh-harrier-habitat/
[160] https://www.cityam.com/totally-without-merit-judge-rejects-environmental-challenge-against-sizewell-c-nuclear-plant/
[161] https://www.statista.com/statistics/426157/united-kingdom-uk-attitudes-towards-nuclear-energy/
[162] https://www.britainremade.co.uk/fast_reliable_public_transport
[163] https://foreignpolicy.com/2022/07/16/immigration-us-technology-companies-work-visas-china-talent-competition-universities/
[164] https://www.mercatus.org/students/research/working-papers/regulation-and-cost-child-care
[165] https://www.bmj.com/content/349/bmj.g7083
[166] https://pris.iaea.org/PRIS/WorldStatistics/LifeTimeUnplannedCapabilityLossFactor.aspx
[167] https://www.nytimes.com/2024/02/22/business/uk-nuclear-power.html
[168] https://www.ncsl.org/research/environment-and-natural-resources/states-restrictions-on-new-nuclear-power-facility.aspx

[169] https://hal.science/hal-00956292/document
[170] https://www.iaea.org/newscenter/news/towards-a-just-energy-transition-nuclear-power-boasts-best-paid-jobs-in-clean-energy-sector
[171] https://www.britishscienceassociation.org/Handlers/Download.ashx
[172] https://www.orano.group/en/2021-misconceptions/41-of-the-french-public-think-that-nuclear-energy-is-a-sector-that-creates-jobs-in-france
[173] https://www.ornl.gov/news/3d-printed-nuclear-reactor-promises-faster-more-economical-path-nuclear-energy
[174] https://www.freethink.com/opinion/nuclear-power-coal
[175] https://www.iaea.org/publications/15198/artificial-intelligence-for-accelerating-nuclear-applications-science-and-technology
[176] https://www.theguardian.com/environment/2010/jun/14/police-compensation-kingsnorth-climate-protesters
[177] https://www3.uwsp.edu/cnr-ap/KEEP/Documents/Activities/Energy%20Fact%20Sheets/FactsAboutOil.pdf
[178] https://www.ft.com/content/6cad1883-f87a-471d-9688-c1a3c5a0b7dc
[179] https://www.ft.com/content/6cad1883-f87a-471d-9688-c1a3c5a0b7dc
[180] https://www.smithsonianmag.com/innovation/how-lego-is-constructing-the-next-generation-of-engineers-37671528/
[181] https://humanprogress.org/what-might-an-energy-rich-future-look-like/
[182] https://www.youtube.com/watch?v=wfowl73cDhY
[183] https://insideclimatenews.org/news/27022024/federal-data-reveals-surprising-drop-in-renewable-power-in-2023/
[184] https://www.nature.com/articles/d41586-021-01020-z
[185] https://www.newscientist.com/article/2248729-extinction-rebellion-were-not-veteran-protesters-new-analysis-shows/
[186] https://www.youtube.com/watch?v=9KjbxbyCrNE

ABOUT THE AUTHOR

Zion Lights is a science communicator who has challenged environmental orthodoxies and helped reshape modern energy discourse. She founded Emergency Reactor, the UK's first pro-nuclear environmental campaign, and has played a key role in shifting global opinion toward a more evidence-based and humane vision of energy and progress. As a former editor and spokesperson for Extinction Rebellion, she pushed for scientific clarity inside the climate movement, speaking publicly, writing widely, and launching *The Hourglass* – the first dedicated climate-reporting newspaper of its kind. Her commitment to evidence-led environmentalism earned her the Holyoake Lecture Medal from Humanists UK, recognising her as a leading voice for a future grounded in innovation, compassion, and human flourishing.

www.zionlights.co.uk

INDEX

E